Simply
CONTAINERS

Simply
CONTAINERS

Bright ideas for your patio, balcony, windows & walls

DEALERFIELD

This edition specially printed for:
Dealerfield Ltd
Glaisdale Parkway
Glaisdale Drive
Nottingham
NG8 4GA

Conceived and produced by Brown Packaging Books Ltd,
255–257 Liverpool Road, London N1 1LX

A British Library Cataloguing in Publication Data block for this book may be obtained from the British Library

ISBN: 0 7063 7646 3

Printed in Italy

Picture credits
All photographs Fabbri Publishing/Robert Harding Syndication except the following:
page 40: S&O Mathews; page 60 (bottom): EWA; page 71: Sonia Baker; page 135 Walter Blom & Son

Contents

Continued over page

Contents .... continued

A BOUNTIFUL MANGER

Once a common sight in stable yards, hay mangers make ideal wall-hung containers for eye-catching Summer displays of brilliantly-coloured annuals.

MATERIALS LIST
You will need: a
wrought-iron wall-
mounted hay manger
(1); a bag of hay or
straw (2); *Phlox
drummondii* (3); large-
flowered Petunias (4);
plastic lining sheet (5);
peat-based potting or
container compost (6);
trowel (7); *Dianthus
'Magic Charms'* (8);
Fuchsias (9).

I n the days when horses were the main form of transport, large metal mangers holding hay were a common sight on the walls of stable yards. Few have survived, although sometimes it is still possible to find old ones in junk or antique shops. However, garden centres are now selling good-quality replica mangers, made of plastic-coated wrought iron, which come in a range of sizes and make ideal wall-hung planting baskets for displays of Summer annuals.

Colourful cascade

Lining mangers with hay or straw adds to their appeal, and they look very effective when planted up with boldly-coloured flowering and cascading plants.

Choose a colour theme for the planting which either harmonizes or contrasts dramatically with the colour of the background wall. Be adventurous in your selection of colours, rather than playing safe with the usual mixture of reds, whites, pinks and blues.

Striking effect

The brilliant cerise and purple of the Fuchsia bells is continued in this scheme with strong red and purple Petunias, deep wine-red Dianthus, and a glorious palette of reds, mauves, pinks, and purples in the dwarf Phlox; seen against the strong blue of the wall, the effect is striking.

Try using other colour combinations, either blending or contrasting with the manger's surroundings, to make a dramatically-colourful feature throughout the Summer.

PICK & PLANT

Other colourful combinations of Summer annuals include:
❑ Yellow and orange schemes with Pot Marigold (*Calendula*), Begonias, *Tagetes*, *Calceolarias* and Nasturtiums.
❑ Bold red displays using *Impatiens*, Begonias, Fuchsias and Petunias.
❑ All-pink collections with Ivy-leaved and Zonal Pelargoniums, Petunias and Snapdragons.
❑ Subtle blues including *Ageratum*, *Campanula* and different shades of Lobelia.

PREPARING THE HAY MANGER

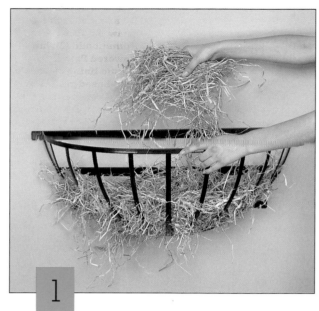

1

To give the hay manger maximum character, line it with a thick layer of either hay or straw, which can be bought from a pet shop. Try to find hay with thick golden-coloured stalks, rather than the thinner green type, which is less visually appealing.

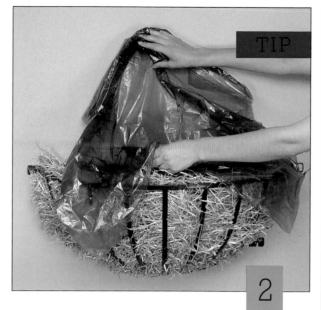

TIP

2

To retain moisture within the planter, and to help keep the layer of hay or straw dry so that it does not rot, place a piece of plastic lining inside the hay. Trim off the excess, or tuck it down inside the hay so that it cannot be seen.

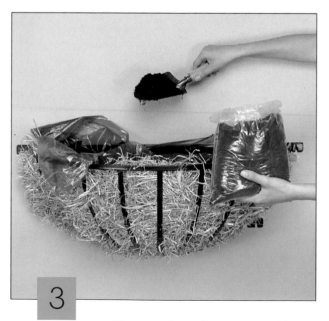

3

Use peat-based compost, which is lighter than the loam-based type, an important consideration when a large planter is to be wall hung. Partially fill the planter with compost, to a level which will enable plants to be positioned at the right height.

4

Water all the plants in their individual pots and allow them to drain. Begin the planting with a group of dark red Dianthus placed centrally at the back of the manger. Keep the plants fairly high and surround them with compost to keep them firm.

COMPLETING THE HAY MANGER

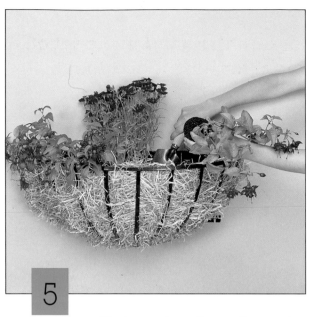

5

Plant a number of cascading cerise and purple Fuchsias around the front rim of the manger, so that they can grow outwards and hang gracefully down the front of the display. Handle these plants with care to avoid damaging them.

6

Plant several brilliantly-coloured dwarf Phlox all around the Dianthus, positioning the different shades to achieve a well-balanced mixture. Carefully fill in the gaps in the compost as planting proceeds, and firm down gently with your fingers.

7

Complete the planting of the hay manger by filling all the remaining spaces with strong red and purple Petunias. Remember that the plants will spread, so do not pack them too closely together, but ensure that the display will look full and generous.

8

Wall-mounted mangers will need watering daily - twice a day in hot weather. Water moderately, but frequently, rather than drenching the container, which will make it very heavy and could cause problems with damp penetrating the wall.

SHINY METAL PLANTERS

Every-day garden utensils are easily transformed into a set of original, shiny planters that are ideal for bright bedding in shades of orange and gold.

PLANTING THE CONTAINERS

MATERIALS LIST
You will need: *Lotus berthelotii* x *maculatus* (1); *Viola* 'Yellow Princess' (2); yellow Antirrhinum (3); Tagetes (4); Gazania (5); *Helichrysum petiolare* (6); *Bidens aurea* (7); galvanized steel watering can (8); galvanized steel bucket (9); galvanized steel trug (10); *Senecio cineraria* 'Silver Dust' (11); compost (12); moisture-retentive clay pellets (13); trowel (14).

With a little imagination, it's easy to conjure up interesting ideas for new and unusual planters. These smart, galvanized steel garden utensils make a welcome change to the normal range of flower pots and tubs. Moreover, they can be picked up at relatively low prices from a variety of sources, such as old-fashioned hardware stores.

The bright silver finish of these containers will cheer up dull corners of the patio and is effectively complemented by many bedding colour schemes. Here, shades of orange and gold are used to contrast with the silver, but various pastel pink and blue shades, with silvery foliage, would look just as good for a more subtle, pretty display.

Added interest
If you are not keen on the bright silver finish, the steel can easily be painted, perhaps black or smart white, using a proprietary outdoor paint suitable for metal. The containers could even be decorated with stencilled patterns for added interest.

1 If possible, make a few drainage holes in the bottom of the bucket, can and trug, using a suitable drill. Alternatively, place an 8cm (3in) layer of moisture-retentive pellets in the bottom of each and smooth out to form an even layer.

2 Most bedding plants require a free-draining compost, so if necessary, add up to a quarter by volume of horticultural grit or sand to the mixture to improve the drainage. Fill the bucket to within 8cm (3in) of the rim and firm lightly.

3 Remove the plants from their pots, and begin by planting the tall-growing varieties, such as the Gazania, in the centre of the bucket (inset) and the smaller varieties, like the Tagetes, around the edges. Aim for a full, balanced arrangement.

4 Fill the spaces between the plants with more compost, firming it lightly, then water the plants well, using a watering can fitted with a fine rose. Plant-up the galvanized steel trug and watering can in the same way, using the remaining plants.

CLASSICAL ROOTS

Add a touch of drama to the patio with a classical-style
wall-planter depicting the Greek God Zeus, his hair a writhing
mass of variegated Ivy and sleek strands of black foliage.

PLANTING THE HEAD

MATERIALS LIST
Terracotta head-shaped planter (1);
moisture-retentive pebbles (2);
crocks and potting compost (3); Ivy
(4); *Ophiopogon* (5).

Garden ornaments and statues evoke an atmosphere steeped in history. Terracotta or stone-look planters modelled on figures from mythology create an air of noble decay that suits the formal patio or courtyard garden.

This wall-planter, depicting the chiselled features of Greek god Zeus, has been planted with a magnificent head of 'hair' created by the trailing strands of variegated Ivy *(Hedera)*, its white, green and yellow tones balanced by an arching grassy-leaved clump of stark black *Ophiopogon*. Plant the Ivy so that it trails around the sides of the head, with the central clump of *Ophiopogon* (the blades of which reach about 20cm/8in long) giving body to the display, while forming a neat fringe.

PRACTICAL POINTER

❏ **Fixing the planter** The planter has moulded hole at back for hanging on a wall. Hammer a 10cm (4in) long masonry nail into the mortar joint in brickwork. Slot on planter and adjust its level.

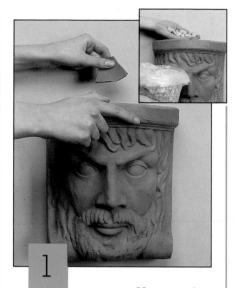

1

Hang up the container on a nail, then place a few crocks in the base, but do not block the drainage hole. Add a handful of moisture-retentive pellets to aid drainge (inset).

2

Use a hand trowel to add potting-and-container compost to the wall head; press the compost down lightly without compacting it. Leave a gap at the top for the plants.

3

Remove *Ophipogon* from its pot and gently tease out its rootball. Place the plant in the centre of the container with its strands overhanging the front; back-fill around it with more compost.

4

Separate the Ivy into small clumps and plant at front rim of head, so that strands will cascade down the sides of the face. Water well before hanging the container in its final location.

STARRY MARGUERITES AND PANSIES

A standard Marguerite, covered with a mass of starry white flowers, will bloom happily through the Summer accompanied by dainty Pinks and the sunny faces of golden Pansies.

MATERIALS LIST
You will need: a large terracotta planter (1); standard Marguerite, *Argyranthemum frutescens* **(2); potting compost (3); trowel (4); crocks (5); yellow Pansies,** *Viola* **(6); white Pinks,** *Dianthus* **(7).**

The delicate Marguerite can often be bought as a small standard tree. The feathery grey-green foliage will have been clipped and trained into a globe shape, above the slender trunk, and during July and August it will be covered in a mass of dainty white Daisies.

Planted in a large terracotta pot, filled with perennial Pinks and annual Pansies, the Marguerite makes a charming focal point for a patio or doorway. Regular deadheading will ensure a long succession of flowers on all the plants.

Both the Pinks and the Pansies will bloom until September. The Pinks can then be left in the pot until next year, or planted out at the edge of a garden border

PICK & PLANT

❏ **Old-fashioned Pinks** are available in several attractive white varieties including 'Mrs Sinkins', 'Musgrave's Pink' and 'White Ladies'.
❏ **Other yellow annuals** that could be included are Zinnia, Tagetes and the Poached-Egg Flower (*Limnanthes*).

PLANTING THE DISPLAY

Cover the drainage hole of the pot with a handful of crocks so that compost is not washed out when you water the plants. Fill the pot with damp peat-based compost to a level which will allow the plant to be positioned correctly.

Remove the Marguerite from its plastic pot and position it so that the top of the compost is slightly below the rim of the terracotta pot, and the trunk of the bush is vertical. Fill the pot with more compost.

Remove the Pinks from their plastic pots and plant them around the base of the bush, pointing slightly outwards. Take care not to damage the rootball of the Marguerite. Firm around the plants with your hands.

Finally, fill the gaps between the Pinks with the yellow Pansies, firming them in well. Then water the arrangement thoroughly. Deadhead all the plants regularly to ensure a continuous display of flowers.

SUMMER GLORY

A shallow bowl, filled with a mixture of colourful and floriferous Petunias, makes a cheerful and glorious show all through the Summer to brighten even the dullest of days.

PLANTING THE BOWL

MATERIALS LIST
You will need: bowl (1); compost (2); moisture-retentive pellets (3); trowel (4); Sweet Alyssum, *Lobularia maritima* (5); Petunia plants (6); trailing Lobelia (7).

Petunias offer real value for money, being inexpensive, yet producing a profusion of glorious blooms over many weeks during the Summer. With all the interesting colours and shapes available, it's really worth giving these old favourites a try.

Here, Petunias are teamed with purple trailing Lobelia, cascading over the sides of the bowl, and Sweet Alyssum that brings a splash of bright white.

A wide choice

There are countless Petunia cultivars - choose from large or small, single or double, frilly, striped or veined, in shades of pink, purple, red, white, blue and even yellow. The multiflora varieties suit exposed sites, as they are more tolerant of damp, but in sheltered places, experiment with the showy grandiflora and double-flowered types.

Plant-up the bowl at the end of May, when all risk of frost has passed, choosing small, but sturdy, plants that haven't had time to become leggy. These annuals, of course, are easy and less expensive to raise from seed - you also get a wider range of varieties from which to choose.

Petunias need a bright, sunny position to encourage flowering. Water well when the compost surface dries, and feed every ten days, from a month after planting. Remove the dead Petunia heads to promote flowering.

1 Pour a 2.5cm (1in) deep layer of moisture-retentive pellets into the bottom of the bowl (inset); then fill the bowl almost to the rim with a good-quality, loam-based compost. Firm the compost with your fingers.

2 Split-up the strips of plants, gently separating the individual plants (or clumps in the case of the Lobelia). Tease the roots apart, making sure each plant is accompanied by a good portion of roots.

3 Plant the Petunia plants quite close together for a really full effect, but not so close that they do not have room to grow. Make a hole in the compost for each, lower the roots in, and firm compost back around them.

4 Use the Alyssum and Lobelia plants to fill between the Petunias, particularly around the edges of the bowl so that the Lobelia can trail over the sides (inset). Firm the compost, adding more if necessary.

BEAN BONANZA

These handsome Climbing French Beans, grown in a bold, painted half barrel, will produce a good crop of delicious crisp beans with a flavour unlike that of supermarket equivalents.

F rench Beans have a delicious, distinctive flavour quite unlike that of Runner Beans. Low in calories and high in vitamins and minerals, these tasty vegetables are ideal for chopping raw into salads or lightly steaming as an accompaniment to many dishes. Home-grown beans have a crispness and fresh taste not found in shop-bought equivalents.

As well as producing lovely beans all Summer, Climbing Bean plants are very attractive, especially when grown in a handsome painted barrel. Their lush green foliage is set off by the clusters of white, pink or scarlet blooms produced throughout the season.

There are Dwarf as well as Climbing varieties of French Beans: the Climbing varieties produce a fine crop of pods over a longer period than the Dwarf varieties, and you don't need to bend down to harvest them.

Sow the beans indoors in April and keep them in a warm place until they germinate. When the young seedlings emerge, move the pots to a bright, cool place where they can grow on for about four

MATERIALS LIST
You will need: Above, half barrel (1); cane (2); hammer (3); medium-sized paintbrush (4); small paintbrush (5); knife (6); crocks (7); scissors (8); trowel (9); garden twine (10); staples (11); wood primer (12); wood preservative (13); black metal paint (14); white exterior gloss (15); Below, compost (16); seed tray (17); peat pots (18); ruler (19); pencil (20); name tags (21); Climbing French Bean Seeds 'Blue Lake'(22).

weeks before being planted outside into the barrel. Water thoroughly after planting out until well established. Feed the beans regularly, from six weeks after planting, with a potash-rich food, carefully following the manufacturer's instructions. Water generously when the compost surface has begun to dry. Site the barrel in full sun, in a sheltered position, as these plants are vulnerable to wind damage.

Regular picking of the beans keeps the plants producing a good crop. Harvest when young and tender, as mature pods are tougher and show developing beans bulging out of the sides. For the tastiest, crispest beans, harvest regularly and eat as soon as possible after picking.

PICK & PLANT

Many varieties of Climbing French Bean seeds are available from seed companies, such as:
❏ **Blue Lake** - this white-seeded variety produces small clusters of little, round, fleshy pods that are tender and stringless.
❏ **Kwintus** - the 25cm (10 in) long pods have seed bulges along the whole length. A distinctive flavour and usually stringless.
❏ **Kentucky Wonder** - this vigorous grower produces magnificently-flavoured, straight, fleshy pods borne in clusters for easy picking.
❏ **Musica** - an early, high-yield of wide, flattened pods with a real beany flavour.
❏ **Purple Podded** - this decorative bean has deep purple pods and a delicious flavour.
❏ **Hunter** - a strong-growing variety producing a heavy crop over a long period. The flat, stringless pods grow to 22.5cm (9 in) long.

PREPARING THE SEEDS

1

Separate the peat pots and arrange in a seed tray; these pots are ideal for cultivating large seeds, as seedlings can be transplanted into the barrel still in the pots, which gradually break down, avoiding root disturbance.

2

Crumble some peat-based general multi-purpose compost in to each peat pot, breaking up lumps between fingers. Make sure the compost gets into the bottom and corners of the pots and fill to the rim.

3

Strike off excess compost with a ruler so the pots are not too full (inset); empty loose compost from the tray. Make a hole in the centre of each pot using your index finger; push down 5cm (2in), twisting to compact the sides.

4

After soaking the seeds in warm water overnight, drop one into the bottom of each hole, then gently fill the hole with your fingertips, firming the surface (inset). Water well, using a watering can fitted with a fine rose.

PREPARING THE TUB

5

The barrel must be treated with a wood preservative to prevent it rotting, and to prevent the paint being soaked off by water leaching out from the compost inside. Follow the manufacturer's instructions and apply the preservative in an even coating all over the wood surface.

6

When the preservative has dried thoroughly, a wood primer can be applied. A white primer must be used under white gloss paint. Using the larger paintbrush, apply the primer according to the instructions on the tin, taking care not to get any on the metal bands.

7

Apply the exterior gloss in the same way after the primer has dried. Use firm, even strokes with the same brush, after having cleaned it thoroughly. You will need to apply two coats of gloss to get a bright, white finish with no dark areas of wood showing through.

8

When the white gloss is dry to the touch, you can blacken the metal bands with metal paint. Use the small paintbrush and apply the paint sparingly to stop any runs or drips getting on to the white wood. Wipe off any drips quickly with a soft cloth and apply two coats if necessary.

9

If there are no drainage holes in the bottom of the barrel, turn it upside down and drill about three holes, 3cm (1in) in diameter, with a wood bit. Put a handful of large crocks, or pieces of broken flower pot, over each drainage hole to prevent compost leaching during watering.

10

Fill the barrel with compost, taking care not to dislodge the crocks in the process. Firm the compost in the bottom of the barrel, making sure it fills all the nooks and crannies, and keep on firming all the way up until the barrel is full (up to 5cm/2in below the rim).

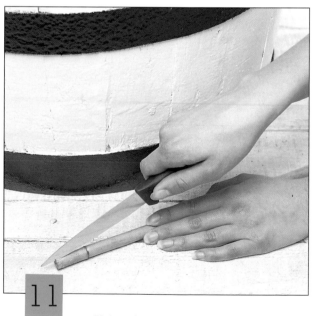

11

Using the sharp knife, cut a slit in the top of the cane right the way through and about 2.5cm (1in) long. Twist the cane and cut another slit at right angles to the first so that when the cane is viewed from the end, the cuts form an X-shape. These will be used to support the strings.

TIP

12

Push the length of cane in to the compost in the centre of the barrel, with the cut end uppermost, and keep pushing it until the cane has reached the bottom. You will need a cane about 2.1m (7ft) long so there will be about 1.5m (5ft) above compost level to support the plants.

PLANTING THE BEANS

13 With a large hammer, knock eight strong staples into the top edge of the barrel, to make four opposite pairs through which the string will be threaded. Knock the staples into the middle of the rim so that they don't split the wood. Leave half an inch of staple protruding from the wood.

TIP

14 Cut off four lengths of twine, each about 3.3m (11ft) long. Tie the end of one of these securely to one of the staples, take the piece up over the cane (through a slit in the top) and tie the other end to the opposite staple. Repeat this with the other three pieces of twine and pairs of staples.

15 Transplant the Climbing French Bean seedlings in to the barrel when they are about four weeks old. Plant them, still in the peat pots, about 10cm (4in) apart around the edge of the tub, 10cm (4in) from the sides. Wind any long shoots on the plants around the strings if they will reach.

16 When the seedlings are in place, use some lengths of twine to make horizontal supports for the beans. Tie the pieces in a circle around the vertical strings, twisted around each string to keep them in place. Tie the strings quite taut and repeat this three or four times equally spaced.

CHEERY CHIMNEY POTS

Salvaged chimney pots make splendid planters for masses of
summer blooms such as geraniums. Group together a selection
of sizes for a colourful patio display.

MATERIALS LIST
To assemble the chimney pot planter display you will need: a selection of chimney pots, such as this clay crown pot (1); a number of plastic plant pots and crocks (2); container compost (3); electric drill for making holes in the pots (4); tape measure for measuring up pot sizes (5); butcher's hooks (6); and plants such as ivies (7); and, below and above respectively, trailing and upright geraniums (8).

Redundant chimney pots, salvaged from their lofty positions due to demolition of an old building, can take on a new role as patio planters.

Group together a selection of pots in a sunny corner of the patio and plant out with summer blooms such as geranium, pelargonium, fuschia, busy lizzie, interspersed with trailing ivies to hang down from the rim.

Sizes and styles

Chimney pots come in a diverse range of sizes and styles, in clay or less attractive concrete. You will probably be surprised at how tall and heavy they are once removed from their stacks: when seen from ground level pots appear small yet can be anything up to about 1.2m (4ft) tall.

Designs can be conical, with a rounded rim; shorter and more stubby, with broad base and squared rim; or capped with a pointed crown. Some pots are pierced around the neck with slots that improve the draw of the fire, which can be utilized by inserting plants such as trailing geraniums or delicate lobelia.

Fixing the plant pots

With such tall planters, it is not practical to fill them with compost, as you would with a conventional tub: not only would this planting medium be too deep and unstable for the plants to survive adequately but also it's a method wasteful of costly compost. Set the plants in ordinary plastic pots suspended from the rim of the chimney pots, or wedge them into the mouth of the pot.

FIXING FACT

To suspend a plastic plant pot from the rim of a chimney pot, first drill holes through the side of the pot into which you can insert buter's hooks or home-made wire hooks. Drill from inside of the pot into an offcut of wood to avoid splitting the plastic.

PLANTING THE SMALL CHIMNEY POT

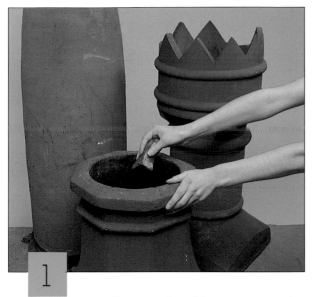

1

Arrange the chimney pots on the patio as required, aiming for a pleasing balance of shapes and styles. The smallest pots can be planted as a conventional container. Place pieces of broken bricks into the base of the pot to assure adequate drainage.

2

Trowel loam-based compost, such as John Innes No.3, into the smallest chimney pot, covering the drainage crocks, and filling it to within about 15cm (6in) of the top rim. As you work, gently firm up the compost with your hands, but avoid compacting it too much.

TIP

3

Plant the geraniums - still in their plastic plant pots - in the top of the smallest chimney pot. Leaving the plants in their pots enables you to turn them to avoid lopsided growth, and makes it easier to remove them at the end of the season for overwintering.

4

Arrange the pots of upright geraniums in the top of the chimney pot, then carefully part the stems and trowel more compost around the pots, firming it with your hands. You may at this stage wish to introduce trailing ivies around the perimeter of the pot planter.

SUSPENDING THE INNER POTS

5

Measure the diameter of the other chimney pots to enable you to determine the size of plastic plant pot you will need to fit into it. Some chimney pots incorporate an inner ledge on which the plant pot can sit. Check that the pot will sit snugly before planting (inset).

6

Place crocks in the pot inserted in the top of the chimney pot, then fill with compost. Remove the geraniums from their existing pots, gently tease out the roots from the root ball, then plant in the plastic pot. A mix of upright and trailing plants gives the fullest display.

7

Plant clumps of ivy between the geraniums, so that the stems will trail down the sides of the chimney pot, between the points of the 'crown'. By fixing the plant pot lower down in the chimney you may be able to introduce trailing plants to the slots in the chimney pot.

TIP

8

If you need to suspend a smaller plant pot in a plainer chimney pot, attach butcher's hooks in holes pre-drilled in the side of the plastic pot (see Fixing Fact), and hang the pot from the rim. Separate strands of ivy and plant as before (inset), then add geraniums.

HIDDEN ASSETS

Paved areas near the house are often marred by unsightly manhole covers - disguise such an eyesore with a low-level planter filled with shade-loving plants.

MATERIALS LIST
You will need:
peat-based potting
compost (1); a
shallow wooden
planter (2);
moisture-retentive
pellets (3); scissors
(4); plastic lining
(5); trowel (6); Wax
Begonia, *Begonia
semperflorens*
'Devil White' (7);
Hosta 'Royal
Standard' (8);
Buckler Fern,
*Dryopteris
pseudomas* (9);
Hosta fortunei
'Obscura
Marginata' (10);
Hosta fortunei
'Albopicta' (11).

So often, terraces, patios and other areas of paving around a house are marred by unattractive manhole covers over drains; this need not be a problem, as they can be easily covered and disguised with shallow timber planters, which are sold for just such a purpose. If the planter is placed in a sunny position, it can be filled with brightly-coloured annuals; if, however, the position is a shady one, as is often the case, the plants selected can be a little more unusual, to take advantage of the shade.

Contrasting foliage

Both perennial Ferns and Hostas will thrive in shade, and their attractively-contrasting foliage will remain refreshingly green throughout the Summer. Additional colour and interest can be provided by including the easily-grown annual *Begonia semperflorens* 'Devil White'. The Begonia's glossy, bronzed foliage around the margins of the display will form a visual link with the dark-stained timber of the planter, and the waxy white flowers will be borne in profusion until the arrival of the first frosts in the Autumn.

Bold and decorative

Hostas are handsome plants, grown for their bold decorative foliage, rather than for their flower spikes of delicate lilac or white. There is a considerable range of variation in the leaves: from the blue-green ribbed and quilted foliage of *H. sieboldiana*, through all the interesting combinations of variegation - like the wavy dark green leaves of *H. crispula* edged with white, or the yellow-splashed foliage of *H. fortunei* 'Albopicta'. Plants have even been developed with pure golden leaves.

When choosing plants for this type of display, try to achieve pleasing contrasts of size, colour and leaf shape.

Essential moisture

Like Ferns, Hostas will perform best in rich, peat-based compost. What they need most is moisture - the planter must be kept damp to ensure that the plants grow happily. This is particularly important if the shady site is created by a large tree, which will prevent all but the heaviest rain from reaching the soil around the plants.

PRACTICAL POINTERS

❑ **Position** The completed planter will be heavy, so place it in its intended final position before starting to plant it up. To ensure good drainage, place bricks underneath the planter to raise it above the paving.

❑ **Preservation** If you buy an untreated wooden trough, or make one yourself, protect it by applying 2-3 coats of timber preservative; there are many colours and types to choose from, but make sure that you buy the type that is not harmful to plants.

PREPARING THE TROUGH

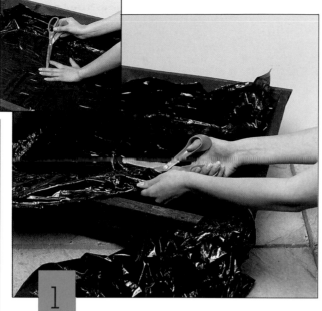

Most wooden planting troughs should already be treated with a plant-friendly timber preservative (if not, see Practical Pointer). However, to ensure that the planter lasts for as long as possible, line it with plastic and pierce some drainage holes using the scissors (inset).

To help retain the maximum amount of moisture in the compost, which is essential for all the plants to grow well, place a layer of moisture-retentive pellets, about 4cm (1 3/4in) deep, in the bottom of the planter. The pellets will absorb water and release it as needed to the plants.

Place a deep layer of peat-based potting compost in the planter, adding sufficient to bring the depth to a level that will allow the larger plants to be planted easily at the correct height. Water the compost and the plants well before planting, and allow time for the water to soak in.

Plan the display, depending on whether it will be seen from one side only, or all round. The tall Ferns should be used at the back or centrally, with the larger Hostas surrounding them, and the smaller Hostas and Begonias grouped around the edge.

PLANTING THE TROUGH

5

Plant the larger *Hosta* 'Royal Standard' around the tall Ferns to form a dense area of bold blue-green foliage. If the plants are difficult to remove from their plastic pots, tap all around the pots with a trowel to release the rootballs and prevent damage to the root systems.

6

Continue planting the medium-sized Hostas around the larger plants, aiming to create a pleasing contrast of leaf colours, patterns and sizes. Then position the lighter *H. fortunei* 'Albopicta' plants to cascade over the rim of the planter, helping to soften its angular outline.

7

Complete the planting by placing the small, bronze-leaved *Begonia semperflorens* around the rim of the planter. Avoid a regimented row of plants - they should be grouped informally among the smaller Hostas, linking with the colour of the dark-stained timber.

8

Fill any gaps between the plants with compost and firm down well. Water the display again, and keep watered daily. Hostas are a favourite with slugs, so if your garden is troubled by them, apply slug pellets or a liquid deterrent, which will be safer with children and pets around.

A SHADY SPOT

Add interest to a shady corner of a courtyard garden throughout the Summer by filling a terracotta herb pot with assorted small Ferns and pure white Busy Lizzies.

MATERIALS LIST
You will need: a terracotta herb pot
(1); Japanese Painted Fern, *Athyrium
niponicum* 'Pictum' (2); White Busy
Lizzies, *Impatiens* (3); Crested Harts
Tongue Ferns, *Asplenium
scolopendrium* (4); potting compost
(5); a plant pot saucer (6); trowel (7);
hammer (8); coarse bark or peat (9);
broken flower pots (10); *Adiantum
pedatum* var. *subpumilum* (11);
Miniature Lady Fern, *Athyrium filix-
femina* 'Minutissimum' (12); *Athyrium
filix-femina* 'Frizelliae' (13).

PICK & PLANT

❏ White Busy Lizzies will
show up clearly in a shady
corner and will contrast
well with the colour of the
terracotta. You could use
salmon pink, or a dazzling
orange and white striped
variety instead.
❏ Any hardy Ferns could
be used, even the larger
growing varieties will stay
small enough to remain in
the pot for one season.

Shady areas of a courtyard
can be just as attractive as
more open, sunny spots,
provided you choose plants that
will be happy in dull places.
Busy Lizzies (*Impatiens*) grow
very quickly, will begin to bloom
at the start of the Summer and,
given regular watering and
feeding, will go on flowering until
the first frost of Autumn. Combine
them with shade-loving Ferns
whose attractive foliage will
provide Summer green in paved
or walled areas where there is
no room for a lawn or trees.

Pot sizes
Herb pots - which can also be
used for growing strawberries -
come in several sizes, the
medium size, with eight planting
holes in the sides, was used for
this display. Choose the size of
pot that best complements the
space available, and adjust the
number of plants used accordingly.

Selecting the plants
Chose mature Busy Lizzies that
have reached flowering size in
their pots. The best Ferns are
those sold in 6cm (2in) round
pots, often called 'tots', as their
rootballs will slip easily through
the holes in the sides of the pot.
If these are not available use
plants in slightly larger pots,
divide the rootballs with a sharp
knife and use half in each hole.

Positioning
Position the pot where it can be
viewed from all sides, and turn it
round every three or four days
to ensure even growth. Both
Busy Lizzies and Ferns prefer
moist conditions, so stand the pot
in a matching saucer and keep it
topped up with water.

Over - wintering
The Busy Lizzies will be killed by
the first frost, but Ferns are
perennials and should survive
the Winter. However, because
plants in containers are more
vulnerable to frost than those
growing in the ground, it is wise
to move the pot under cover in
severe weather.

Prolonging the display
The Ferns will provide some
interest throughout the Winter
but, to keep the display going,
remove the Busy Lizzies after
they have been frosted and
replace with Primroses. Use a
hardy outdoor variety and plant
them around the rim of the pot,
together with a group of early
Narcissi bulbs packed closely
together in the centre, to provide
colour right through to the
following Spring.

PREPARING THE POT

Place a layer of coarse material at least 2.5cm (1in) deep in the bottom of the pot to assist drainage. Break some old clay flower pots into pieces roughly 2.5cm (1in) square with a hammer. Alternatively, if these are not available you could also use pieces of old brick.

The layer of crocks should go into the the pot with the largest pieces first topped by the smaller ones. Further assist drainage by covering them with a layer of a similar depth of coarse bark, or bits of peat sieved out from a bale when last making up potting compost.

Fill the pot with enough potting compost to bring the level up to the bottom of the lower row of holes; press the compost down so that it is fairly firm. Use compost that is just damp, but not moist enough to stick to the fingers. You may need to stir in some water to achieve this.

To remove a Fern from its pot spread the fingers of your left hand (right hand if you are left-handed) over the surface of the compost, tip the pot upside down and knock the pot rim on the edge of a hard surface. The plant should then slip out of the pot with its rootball intact.

PLANTING THE FERNS

5

Use two different kinds of Ferns
alternately in the four lower holes to
give a contrast of foliage. Take a
plant in one hand and push its
rootball in through one of the holes.
Use the other hand to press the
rootball firmly into the compost so
that the foliage is outside the pot.

6

Plant the other Ferns in the same way,
adding more potting compost to
bring the level up to the bottom of the
top row of holes. Firm the compost
well around the rootball of each
Fern. To avoid accidental spillages of
compost, stand the bucket close to
the pot for easy access.

7

Plant the second row of Ferns in the
same way as the first row, again with
the two kinds alternating to provide a
good contrast between the foliage in
the top and bottom rows of holes.
After planting keep adding more
compost until the pot is filled up to
near the top of the rim.

8

Remove the *Impatiens* from their pots
using the same method employed for
the Ferns. Using a trowel, plant them
in the top of the herb pot, evenly
spaced from one another and about
4cm (1 1/2in) in from the sides. Keep
the rootballs intact and firm compost
around them, adding extra if needed.

GLORIOUS CAMELLIAS

With their showy blooms and glossy foliage, Camellias make superb plants to brighten up a patio or a cool conservatory, providing a rich choice of colour and variety.

MATERIALS LIST
You will need: *Camellia japonica* 'Haku-raku-ten' (1); black plastic sheeting (2); a trowel (3); moisture-retentive clay pellets (4); scissors (5); a Versailles tub (6); slightly acidic compost (7).

Named after Georg Josef Kamel (a 17th century Jesuit who studied the flora of the Philippines), the Camellia is a large and attractive shrub of the *Theaceae* family from tropical Asia, and the 80 or so species of trees and shrubs that make up the genus can be found everywhere between India and Japan. Some have large blooms, while others are grown for their lush and glossy green foliage. Camellias make superb plants for an unheated conservatory and are an excellent choice for growing in containers.

Manageable height

The plant shown here is the *Camellia japonica* 'Haku-raku-ten', which is a semi-double form (see Pick & Plant overleaf). If grown outside, it will turn into a substantial tree, eventually reaching a height of 9m (30ft). However, if containerized, the plant will only grow to a more manageable height of about 3m

(10ft). The eventual height will be dependent on the size of the pot that is used and how much the roots of the plant are constricted.

Elegant lines

Because the *Camellia japonica* has such a graceful habit, it is important to select its container with care. A Versailles tub, as shown here, with its elegant lines and warm, rich colour, will complement the dark and shiny green of the foliage, making this an attractive and spectacular display that requires very little by way of care and attention, but one that will give an enormous amount of pleasure.

Water regularly

For the best results, *Camellia japonica* requires a minimum temperature range of 5-10°C (40-50°F). Position it in full sun or partial shade, and ensure that it is watered regularly; never allow the surface of the compost to dry out completely.

FIXING FACTS

❏ **Assembling the tub**
Versailles tubs are available in a variety of forms and in various finishes. This particular wooden example comes as a kit for home assembly. All the parts fit together very neatly, and the tub can be put together in a matter of minutes, using the socket-headed screws provided - the kit also includes an Allen key for tightening the screws. The tub then forms a very sturdy planter, which will stay looking attractive for years to come, enhancing the appearance of any plant you choose to display in it.

Assemble the sides and legs of the Versailles tub, making sure the screws are tight.

PREPARING THE TUB

Line the base of the Versailles tub with plastic sheeting, piercing holes in it with scissors to prevent the compost from becoming waterlogged (inset). On top of the sheeting, trowel in a 5-10cm (2-4in) layer of moisture-retentive clay pellets to aid the drainage of the compost.

Make sure that the tub is in the position chosen for the display, as it will be very heavy and difficult to move once it is filled with compost. Then trowel in a neutral pH or slightly acidic, potting medium until the Versailles tub is filled to within about 20cm (8in) of the rim.

Gently remove the plant from its pot by resting it against the side of the tub and tapping gently with a trowel, rotating the pot as you do so. This will loosen the grip of the compost on the pot, allowing the rootball to be removed without causing any damage to the root system.

Place the Camellia in the tub and fill around it with the potting medium, firming it down by hand as you go; remember to over-fill the pot slightly to allow for the inevitable settlement of the compost. Water the plant well and keep well watered (inset). Do not allow it to dry out.

Above: *Camellia japonica* **'Lady Clare'** - a large-flowered, attractive peach-pink Semi-Double form with white stamens.

Below: *Camellia japonica* **'Rubescens Major'** is a bushy, Double-flowered plant with crimson blooms and dark veins.

There is a wide range of varieties of *Camellia japonica* to choose from, so whatever your tastes, you should be able to find one that will be right for the situation you have in mind. In addition to the selection of colours, the cultivars are available in a variety of flower shapes, which are classified by the arrangement of the petals and stamens:

❑ **Single-flowered** have eight or fewer petals.

❑ **Semi-double** - at least two rows of petals with a conspicuous central boss of stamens.

❑ **Anemone-form** - one or two rows of petals with a central boss comprising a mass of petalodes (leaves or stamens that take the role of petals) and stamens.

❑ **Paeony-form** - having no regular form and being full-petalled with petaloids and a few stamens.

❑ **Double** have overlapping petals, the stamens being in a concave centre.

❑ **Formal Double** have many overlapping petals in regular rows and no visible stamens.

Above: *Camellia japonica* **'Contessa Lavinia Maggi'** is a delicate creamy-pink Formal Double with dark rose-coloured stripes.

Below: *Camellia japonica* **'Eximea'** is a large-flowered Formal Double with star-shaped blooms and marbling of the petals.

SCREEN BEAUTY

A trellis panel, cloaked in an evergreen climber, will screen a blank wall or an ugly view; with the addition of a few hanging pots, it will make an attractive feature in its own right.

MATERIALS LIST

You will need: *Pelargonium* (1); *Helianthemum* 'Double Yellow' (2); *Hedera* 'Sulphur Heart' (3); **plastic flowerpot (4); clay hanging pots (5); perlite (6); potting compost (7); horticultural grit (8).**

Traditionally, trellis is used to support climbing plants, but by hanging pots from it as well, you can turn it into a truly versatile and attractive feature. Grow an evergreen climber on the trellis to provide a backdrop for pots of bright flowers or contrasting foliage. Use colourful annuals, or perhaps evergreen perennials for all-year-round interest.

Make sure that the trellis panel is firmly attached to a wall or posts to support the extra weight of the pots. These should be securely wired to the trellis panel or hung on sturdy, S-shaped butchers' hooks.

FIXING FACT

❑ **Fixing trellis to walls**
Plants need room to climb freely around trellis, so use cotton reels or wooden blocks as spacers to hold the panel clear of the wall.

1

If the hanging pots don't have flat bases, stand them in a flowerpot while you plant them up. Since the pots have no drainage holes, add a layer of drainage material, such as perlite (inset). Then put in some compost.

2

Remove the *Pelargonium* from its plastic pot by holding it upside down and squeezing the pot to release the rootball. Place in the clay pot and fill around the rootball with more compost, firming with your fingertips.

3

The *Helianthemum* is a rock plant and prefers a free-draining compost, so mix some horticultural grit with the compost before planting up the hanging pot. Make sure the grit and compost are mixed thoroughly.

4

Remove the *Helianthemum* from its pot and stand it on the compost in the clay pot. Make sure the top of the rootball is just below the rim and fill around it with compost mix. Firm lightly. Pot-up the Ivy in the same manner.

FLOURISHING FOLIAGE

A clever combination of pleasing scents, boldly variegated leaves, fine-cut foliage and bright blooms ensures this compact terracotta pot will provide year-round interest on the patio.

PLANTING THE DISPLAY

MATERIALS LIST
You will need: terracotta pot (1);
Lavender Cotton, *Santolina*
chamaecyparissus **(2); Dead Nettle,**
Lamium maculatum **(3); Winter**
Creeper, *Euonymus fortunei*
'Emerald and Gold' (4); *Artemisia*
schmidtiana **'Nana' (5); Variegated**
Thyme, *Thymus vulgaris* **'Variegatus'**
(6); Curry Plant, *Helichrysum*
italicum **(7); Moss Phlox,** *Phlox*
subulata **(8); Creeping Thyme,**
Thymus praecox arcticus **(9); compost**
(10); trowel (11); crocks (12).

S o many planting groups
chosen for use in pots offer
interest for only one
season; for the remainder of the
year, the tub is left empty if the
plants die down for the Winter,
or uninteresting if the plants have
been chosen for flowers alone.

Here, silver Lavender Cotton
and Curry Plant offer Winter
colour, as do the variegated Winter
Creeper, feathery Artemisia, and
dainty Thymes. Moss Phlox
produces masses of star-shaped
blooms in pink or white in early
Summer, Dead Nettle is topped
with spikes of pale purple, and the
Thymes have pink flowers on the
stem ends. In contrast, Lavender
Cotton, Artemisia and Curry Plant
will be covered in bright yellow
blooms during Summer. In
addition, many of the plants have
scented foliage, so if the pot is
placed in a prominent position,
passers-by will brush the leaves
and release a heady fragrance.

A plain terracotta pot is ideal;
look for one which is wide and
low. The faster-growing plants
must be pruned regularly to
keep them from outgrowing the
pot and to keep the foliage
young and lush. Place in a
warm, sunny position; water well
in hot weather, allowing the
surface of the compost to dry
between waterings.

1 Place a handful of
crocks in the bottom of
the pot, ensuring you
cover the drainage
hole (inset). Fill the
pot, up to 5cm (2in)
below the rim, with a
free-draining compost,
adding up to a third by
volume of horticultural
grit, if necessary.

2 Arrange the plants, still
in their pots, on top of
the compost in the pot
to decide on planting
positions. Place the
more erect plants
towards the centre of
the pot, allowing the
trailing plants to be
positioned around the
edges, softening them.

3 Remove the plants
from their pots by
turning them upside
down and tapping the
pot on a firm surface to
release the rootball.
Make a hole in the
compost and lower-in
the roots. Firm
compost back around
the roots.

4 When all the plants
have been placed, firm
all over the compost,
adding more if
necessary, until you
have a flat, even
surface. Water-in the
plants until water starts
to seep out of the
drainage hole
underneath.

KITCHEN GARDEN

A glazed stoneware sink, with its broad, shallow design, makes a novel and striking container for a display of sweetly-scented and richly-coloured Tobacco Plants and Sweet William.

PLANTING OUT THE SINK

MATERIALS LIST
You will need: a shallow stoneware sink (1); gravel (2); compost (3); charcoal (4); a trowel (5); Tobacco Plants, *Nicotiana alata* (6); Sweet William, *Dianthus barbatus* (7).

This glazed stoneware sink, set in a sunny place and planted with tall-growing Tobacco Plants (*Nicotiana alata*) and Sweet William (*Dianthus barbatus*), provides an unusual display that is as attractively scented as it is brightly coloured.

Sweet scents

If the plants are kept in a bright, sunny site, the *Nicotiana* flowers will open in the evenings, and the sweet scent that they produce will carry for some distance. The pretty little Sweet William also has strongly-scented blooms, so the total effect is quite heady.

This early Summer-flowering display will also grow in light shade. However, the lower light levels will stimulate the *Nicotiana* into keeping its flowers open all day long, and the perfume of the blooms will be less powerful.

PICK & PLANT

❏ **Scent selection** Day-flowering cultivars of the *Nicotiana* are available in a variety of colours, however none of these cultivars produces the rich scent of the evening-flowering variety.

1 Cover the plug hole with a layer of crocks to prevent the soil from leaching away each time the plants are watered (inset). Spread a layer of gravel about 2.5cm (1in) deep across the bottom of the container.

2 Once the gravel has been distributed evenly across the base of the sink, sprinkle in a layer of horticultural charcoal. This will perform the function of keeping the potting medium sweet and fresh for the plants.

3 Over the charcoal, add a layer of the potting medium; *Nicotiana* and Sweet William will grow happily in almost any sort of soil. Keep trowelling in the compost until the sink is filled to within 2.5cm (1in) of the rim.

4 Once the sink has been prepared, add the plants, putting the taller *Nicotiana* at the back. Firm the compost around the plants with your fingers to ensure that it is in contact with the roots (inset) and water-in the display.

COOL CORNERS

From Japan, the False Castor Oil Plant adapts readily to a cool climate and because of its liking for shade, when planted in an urn, makes an excellent plant for a North-facing wall.

MATERIALS LIST
You will need: a large terracotta urn (1); Ivy, *Hedera canariensis 'Gloire de marengo'* (2); compost (3); crocks (4); trowel (5); False Castor Oil Plant, *Fatsia japonica* (6).

The False Castor Oil Plant (*Fatsia japonica*) originates from South-East Asia. More widely known as a houseplant, this glossy ever-green with distinctive palm-like leaves, is also quite at home when grown outside; it's well-suited to a position either partially or fully shaded. This makes it an ideal shrub for an individual planter placed against a North-facing wall, or perhaps in an enclosed courtyard, where sunlight is limited. In order for the plant to reach maximum size - some 2m/6ft - you will have to pot it on into a larger urn.

The Ivy, chosen for its cream variegations and broad leaves, to complement the shape of the *Fatsia*, has a purplish tinge in Winter, matching its stems.

PRACTICAL POINTER

❏ **Planting on site** The urn is likely to be quite heavy and awkward to manoeuvre - and will be even more so once it has been filled with compost. Choose a suitable site then collect the compost and plants and arrange on site to avoid having to lift the urn into place.

1 Line the base of the urn with crocks (broken pot shards) as this will aid the drainage of the compost (inset). Trowel in a layer of the potting medium until the urn is approximately half-full.

2 Carefully remove the *Fatsia* from its pot, causing as little damage to the root system as possible. This can be achieved by gently tapping the side of the plant pot with a trowel.

3 Making sure that the *Fatsia* is placed centrally within the urn, fill around the rootball with compost and gently firm it down. Slightly over-fill the urn to allow for subsequent settlement.

4 Add the Ivy around the edge of the urn. Water well, then allow the surface of the compost to dry out before watering again; move into a conservatory if temperature falls below 10°C (50°F).

A SUMMER BALL

A hanging basket filled with a tumbling mass of the romantically-named Heartsease, or Wild Pansies, will provide a glorious ball of colour all Summer long.

MATERIALS LIST
You will need: a hanging basket (1);
Heartsease, *Viola tricolor* (2);
sphagnum moss (3); black plastic
lining (4); scissors (5); moisture-
retentive pellets (6); a trowel (7);
peat-based potting compost (8).

At one time, the Wild Pansy (*Viola tricolor*) was widely seen growing in corn-fields, where its spreading stems would form a dense mat. It is this characteristic which makes the delightful little flower an ideal subject for a hanging basket.

A floral globe
The *Violas* are carefully planted through the plastic lining of the basket, and as the stems grow, they can be made to make further roots by inserting them through new holes. Eventually the basket will become a globe of tiny flowers and foliage, which will last throughout the Summer, given daily watering.

PRACTICAL POINTERS

❏ **New roots** As the stems lengthen, make new holes in the plastic and pin the stems back into the compost with pieces of wire; they will make new roots if the compost is kept moist.
❏ **Lightweight** To make the basket as light as possible, mix moisture-retentive pellets into the peat-based compost.

TIP

1

Line the base of the basket with damp sphagnum moss, which will hide the plastic lining and improve the appearance of the basket until the Violas grow. Balancing the basket on a flowerpot makes the job easier.

2

Cut a piece of double-thickness, or heavy-duty, plastic to line the basket. Allow enough to provide a generous overlap at the rim of the basket. Fill the basket to the top with peat-based potting compost (inset).

3

Depending on the the positions of the wire struts, use a pair of scissors to pierce a series of holes through the plastic lining, about half-way down the basket (inset). Carefully plant Violas through these holes.

4

Complete the planting of the basket sides, ensuring that the plants are firmly rooted in the compost. Fill the top of the basket with the remaining plants and water in well. Water every day, and feed every ten days.

FOLIAGE-FILLED URN

A large classic-style urn, filled with resilient foliage houseplants, looks splendid standing in a courtyard garden throughout the Summer months.

MATERIALS LIST
You will need: a classic urn (1); hammer (2); scissors (3); trowel (4); potting compost (5); broken flower pots (6); Wandering Jew, *Tradescantia albiflora* 'Albovittata' **(7); a Parlour Palm,** *Chamaedorea elegans* **(8); Asparagus Fern,** *A. densiflorus* 'Sprengeri'**(9); Spider Plants,** *Chlorophytum comosum* **(10); trailing Ivies,** *Hedera* **(11).**

Many foliage pot plants will be happy to have a Summer holiday in the garden, and they can look striking when packed closely together in a large urn on a sheltered patio or courtyard. Plants for the urn can be removed from their pots, but if they are to go back into the house in the Autumn, it is far easier if they remain in the pots, which should be sunk rim-deep into the compost. Their roots will draw up moisture by capillary action through the holes in the pot bottoms.

Choosing an urn
Urns made from stone or other traditional materials can be expensive, but there are some good glassfibre and plastic replicas on the market. Go for an urn with simple lines, rather than one that has ornate patterning, if trailing plants are to be grown over the sides.

A deep urn that holds plenty of compost will need watering and feeding less frequently, while an ample width will allow room for a better display of plants. Make sure the best place has been chosen for the urn before planting, as once that has been done, it will be heavy and difficult to move. Avoid siting it where it could be toppled easily.

Planning the display
Place the urn where it can be viewed from all sides like an all-round floral arrangement, and choose foliage plants for a good blend of shapes, shades and habits. Provided there is shelter from chill winds, they will grow in shade or part-shade where most flowering plants would not be happy. There should be a tall plant in the centre and some trailers to cascade down over the sides.

PICK & PLANT

Suitable plants for an outdoor Summer display in a classic-style urn include:
❏ *Yucca elephantipes* or *Cordyline terminalis* to give height.
❏ **Boston Fern** (*Nephrolepsis exaltata* 'Bostoniensis') for its pretty, lacy leaves.
❏ Scented-leaf Geraniums for leaves that look and smell delightful.
❏ **Swedish Ivy** (*Plectranthus)* for trailing over the rim of the urn.
❏ **Baby's Tears** (*Soleirolia soleirolii*) to thrive in dense shade.

BEGINNING THE DISPLAY

1

Place the urn into position before filling, as it may be too heavy to move afterwards. Make sure that it's level, or water may run off one edge and roots will receive uneven amounts of water. To check, use a length of wood wider than the top of the urn and a spirit level.

2

To assist drainage, put a 2.5cm (1in) layer of coarse material, such as pieces of broken clay flower pot, in the bottom of the urn. The largest pieces go in first, topped by the smaller ones. Cover with a similar layer of coarse bark or bits of peat sieved from a bale (inset).

3

Fill the urn with potting compost to about 8cm (3in) from the top. Trowel compost into the urn and spread out evenly, then firm with the palm of your hand (inset). The tallest plant should go in the centre of the urn - an adult Parlour Palm, about 60cm (2ft) high, is ideal.

TIP

4

Stand the plant in the centre of the urn, pressing its pot into the compost so that the rim is just below that of the urn. Add more compost around the Palm to fill the urn to the brim. Hold the bucket near the rim to avoid spilling compost and making a mess around the urn.

TIP

5

Position the Spider Plants evenly
around the Palm in the centre of the
urn, burying their pots to the same
depth, but tilting them very slightly
towards the urn rim so that the foliage
is well displayed. Make sure each
pot is firmed very gently into the
compost with your fingers.

6

Plant the *Tradescantias* between the
Spider Plants, but slightly further
forward. Then plant the trailing Ivies
at equal distances from one another
around the inside of the urn's rim,
with pots tilted slightly outwards to
encourage the stems to trail down
over the sides (inset).

7

Plant the Asparagus Ferns between
the Ivies, but further back from the
urn's rim, and tilted slightly towards
the sides. Plants in 8cm (3in) pots are
ideal, but larger plants can be used if
first divided into two and repotted.
Trowel in more compost to fill any
gaps between the plants.

8

Make sure that the compost is firm,
but not too compacted; level off to
give a smooth surface. Using
scissors, go over the planted urn
removing any yellowed leaves, ugly
straggling pieces of trailing plant, or
foliage damaged during potting: aim
for a neat symmetrical shape.

FRAGRANT FLOWERS

Beautiful, white, scented Tobacco Plants and blue Salvias are teamed in a Summer display by the front door, offering a bold and inviting welcome to visitors.

MATERIALS LIST

You will need: a white plastic tub (1); compost (2); white Tobacco Plants, *Nicotiana alata* (3); *Salvia farinacea* 'Blue Bedder' (4); trowel (5); gravel (6).

The front doorstep is a very prominent position for a display of plants in a tub, so it is important to choose carefully. Seasonal bedding plants are ideal here, since they will allow the display to be changed several times throughout the year for maximum interest. Furthermore, bedding plants tend to be very free-flowering.

This bold display for the Summer months is created by teaming pure white Tobacco Plants with rich blue Salvias. They are planted in a bright white plastic tub, which not only complements the arrangement, but is inexpensive and easy to move around.

The plants can be bought in late Spring. Choose small, sturdy specimens with healthy green foliage and no flowers, as these will progress faster than larger, more mature specimens.

PRACTICAL POINTER

❑ **Watering** Use a free-draining compost in the container to prevent a soggy mixture from rotting the plants' roots. Water freely when the plants are small to encourage strong growth, but reduce this slightly as they get larger to promote flowering.

1

Pour a 5cm (2in) layer of gravel into the bottom of the tub (inset) to aid drainage. Add compost to about 3cm (1in) below the rim, firm with your fingers and top up with more compost to the original level.

2

Release the Tobacco Plants from their pots by turning each pot upside down, while supporting the plant, and tapping it on a firm surface. Plant around the edge of the pot, firming the compost around the roots.

3

If the Salvias come in strips, split the plants gently to avoid damaging the roots. Tease them apart, leaving a good portion of compost and roots with each. Plant them about 5cm (2in) apart in the remaining area.

TIP

4

Firm the compost around all the plants and between them to form an even surface to the tub. Add more compost if necessary. Water the plants in well, continuing until it starts to run from the drainage holes.

A NEW LEASE OF LIFE

You might think that a pair of old tyres should be consigned to the rubbish dump, but with a little paint and some colourful plants, they can be turned into an eye-catching planter.

MATERIALS LIST
You will need: *Campanula glomerata*
'Superba' (1); *Fuchsia* **'Winston**
Churchill' (2); **Zonal Geranium**
'Misty' (3); **compost** (4); **old tyres** (5);
white paint (6); **Verbena** (7);
Dianthus 'Colour Magician' (8);
Verbena 'Carousel' (9).

Large containers for outdoor displays can be expensive, but with a little imagination, it is surprising what you can achieve, using the most unlikely of items. Here, a pair of old tyres has been used to make an effective raised planter that is filled with Summer colour.

If stood on a paved surface, line the planter with polythene and add a layer of drainage material to prevent water and compost leaching out and discolouring the paving. Otherwise, simply build up the compost on the ground.

PRACTICAL POINTERS

❏ **In isolation** Using old tyres to build a raised planter provides an opportunity to grow plants that will not tolerate the soil in the garden. For example, Rhododendron and Azalea hate limey soils, but they will grow happily in the confines of a tyre planter filled with ericaceous compost.

❏ **Weather-resistant** Paint the tyres with oil-based gloss paint or masonry paint to resist the weather.

1

Wash the tyres thoroughly, scrubbing them clean. Then paint them the colour of your choice (inset). Choose a suitable site for the tyres, bearing in mind the plants you wish to grow in them, then simply set one tyre on top of the other.

2

Fill the centre of the tyre planter with compost, firming it down with your hands as you go. If you have not used a polythene lining, make sure you work the compost into the insides of the tyres to prevent any subsidence later.

3

Continue adding compost until it reaches a level where the plant rootballs can be set on it with their tops level with the rim of the top tyre. Then add the central plant of the display, in this case *Fuchsia* 'Winston Churchill'.

4

Add the remaining plants, spacing them around the Fuchsia to give an informal, pleasing arrangement. Fill between the plants with more compost, firming it with your fingertips to ensure good contact with the plants' roots.

HOME-GROWN

Courgettes grown at home not only taste delicious, but the plants make attractive subjects for tubs, producing vibrant yellow flowers and glossy fruits all Summer long.

MATERIALS LIST
You will need: courgette seeds (1); sieve (2); labels (3); pots (4); seed compost (5); cling film (6); rule (7); half barrel (8); free-draining compost (9); trowel (10); crocks(11); grown-on courgette plants (12).

Growing your own vegetables can be a rewarding and enjoyable pastime, the end product being far superior in taste to supermarket equivalents. You don't need a lot of space in which to grow vegetables; many, such as courgettes, are suited to pot-culture on a patio or balcony. Furthermore, courgettes are attractive plants with large, handsome leaves and vibrant yellow flowers, which can be admired along with the fruit they produce.

There are many varieties of courgette, ranging in colour from dark green, through stripy, to yellow. Golden courgettes look particularly ornamental on the plant and will brighten a stir-fry or casserole, or they can be used as a colourful accompaniment to a meal.

A head start
Courgette seeds are best sown indoors in mid March, as they can be growing-on in the warm when it is still too cold outside; they will produce fruit earlier due to this head start. After germination, put the seedlings in a bright place to encourage sturdy growth. It will take the plants about six weeks to produce three or four true leaves, after which they should be planted outside into the container, but not until all danger of frost has passed.

The courgettes will need a large container, filled with free-draining compost and placed in a warm, sunny position. Water the plants generously and pinch-out the growing tip when each plant has about five leaves; this will encourage bushy growth.

Pollination
After a while, the plants will begin to produce flowers. These are either male or female: the female flowers have a swelling behind them that will eventually turn into the courgette. They rely on insects to transfer pollen between male and female flowers to ensure fruiting.

If flowers are produced early in the season, or the weather is very dull, there may be no insects around and it will be necessary to pollinate the flowers by hand. Pick off a ripe male flower, strip the petals off and press its pollen-bearing centre into the middle of any fully-opened female flowers.

Prolonging the harvest
Feed the courgette plants every two weeks when the fruits begin to swell. A tomato fertilizer is ideal. Cut off the fruits with a knife when they are 10-15cm (4-6in) long. Regular removal of the courgettes from the plants will prolong the harvest.

PICK & PLANT

There are many varieties of courgette seed to choose from; the following are a few examples:
❏ **Gold Rush** - this is very easily grown and produces slim golden-yellow fruits.
❏ **Burpee Golden Zucchini** - this golden variety can be used as a marrow or harvested early as a courgette.
❏ **Ambassador** - a high-yield variety, producing dark green fruits with crisp white flesh.
❏ **Zucchini** - this one produces an early crop of green, cylindrical fruits 15cm (6in), or more, long.

Female courgette flowers have a swelling behind them that grows to form a courgette.

SOWING THE SEEDS

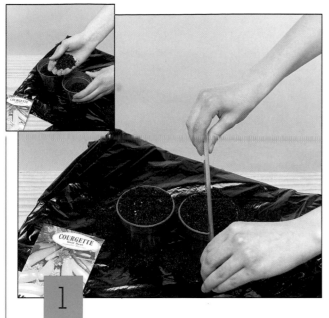

Sieve a few handfuls of compost on to a plastic sheet, using a fine sieve. Use this compost to fill three small pots in which you are going to plant the courgette seeds (inset). Fill the pots to overflowing, and then strike off the excess compost with a rule or similar straight piece of wood.

It is best to put two seeds into each pot, so that if one does not germinate, you will still have the required number of plants. If both seeds come up, weed-out the smaller of the two specimens. Push the seeds, side by side, into the top of the compost.

Compress the surface of the compost lightly with your fingertips to make room for a covering layer of compost over the top of the seeds. Sieve compost into the tops of the pots until it fills them to the rim again. You should aim for a layer about 2cm (3/4in) deep.

TIP

It is helpful to label the pots with the name of the variety and the sowing date, particularly if you are sowing several different batches of seed. Water-in the seeds thoroughly until you see water running from the drainage holes; use a watering can fitted with a fine rose.

PLANTING THE COURGETTES

5

If the pots are covered with cling film, the seeds will not need watering again until after germination. Stand the pots on a shady windowsill and check them every day: as soon as the tiny plants appear, remove the cling film and move the pots to a bright position.

6

When the seedlings are big enough and all risk of frost has passed, they can be planted outside. Add a few handfuls of crocks to the bottom of the barrel to cover the drainage holes and prevent the compost from leaching out each time the courgette plants are watered.

7

Fill the barrel to the top with a free-draining compost. Firm with your fingers and then add more compost until the barrel is full to the brim again. There is no need to make any special provision for the rootballs since the courgette plants are in such small pots.

8

Remove the plants from their pots and arrange them on the compost, making sure that they are equally-spaced and have adequate room to spread (inset). Plant them by making a small hole with the trowel and easing in the rootball. Firm compost back around the roots.

A RED AND WHITE WINDOW BOX

Enhance the view from your window and brighten the exterior of your house with this vibrant display of red and white spring blooms in a terracotta trough.

This terracotta window box has been planted with an array of colourful red and white blooms, which look atractive when viewed either from inside or outside the house.

MATERIALS LIST
Ingredients for the red and white window box are: terracotta clay window box (1); petunias (2); verbena (3); ivy (4), usually several per pot; helichrysum (5); clay pellets for drainage (6); container formula of potting mixture (7).

PICK & PLANT

Plant the container with a display of garden annuals, which will thrive in bright conditions of an exposed outside window sill:

❏ **Verbena** *(Verbena x hybrida)* is a good choice for the taller, back edge of the box: it has long, leafy stems topped with a cluster of tiny blooms.

❏ **Petunia** *(Petunia x hybrida)* Many colours and flower forms of this popular half hardy annual are available (here shown in fetching red and white ringed form), which stands out well against a white-painted sill and window frame, or enhances the often drab appearance of a stone sill.

❏ **Ivy** *(Hedera helix* and forms*)* Fill out gaps around the front and sides of the window box with low-growing creepers such as a variegated form of ivy.

❏ **Helichrysum** *(H. petiolatum)* The silvery, furry-leaved helichrysum is a good choice to crawl over the edges of the box, softening its harsh edges.

A lthough window boxes are ideal if you live in a flat or townhouse - enabling you to grow outdoor varieties of plants where you don't own a garden - they're also an excellent way of brightening a drab outlook. An outside window box filled with a profusion of colourful blooms will not only delicately frame the view beyond but also will soften the exterior face of the house.

Plan the planting arrangement carefully so that the window box has both visual balance and a symmetrical form. Choose a selection of annual plants that will stand the tallest along the back edge of the box, with plants of a more bushy form used for the main part. You can fill in around the front and sides of the box with plants of trailing habit.

Window boxes are available in terracotta clay, stone, concrete or plastic: which you choose depends on personal preference although the clay types retain moisture better than the others, and this is beneficial where the windowsill is exposed to full sunshine.

If you're placing the box on a painted timber sill, or if the wall below the sill is painted, it's best to stand the box on a special tray so that when you water the plants the water will not stream onto the surface and cause staining - or potentially cause the paint to crack or blister.

PREPARING THE WINDOW BOX

TIP

1

Place a few fragments of slate or broken clay plant pot into the base of the window box, covering the drainage holes. These "crocks" will prevent the planting medium from being washed out of the holes during regular watering.

2

Pour a depth of about 3cm (1 1/4in) of absorbent clay pellets into the base of the window box: this will provide good drainage for the roots of the plants without causing silting of the compost, and will also act as a reservoir.

3

Add a proprietary container and potting mixture to the box, covering the pellets. Fill the box to within about 7.5cm (3in) of the top. Settle the mixture gently with your hands, but avoid compacting it too much. Top up with more mixture.

4

Place the larger plants loosely in the box to experiment with the planting arrangement. Line the verbena in a row along the back edge of the box, with the bushier petunias along the front edge. Do not plant the verbena too close together.

PLANTING THE WINDOW BOX

5

When you are happy with the arrangement, remove the plants from their plastic pots and insert in the potting mixture. Tease out any tightly packed roots with your fingers so that they will spread into the potting mixture more easily..

6

Trowel out a hole large enough to take the root ball of the plant. Place the plant in the hole, with its bloom facing the front. Plant the Verbena along the back edge first, then plant the Petunias. Firm the compost around the base of each plant.

7

Separate individual ivy plants from the main clump with your fingers and plant between the Petunias so that their trailing fronds hang down over the front of the window box. Place a single Helichrysum at each end of the window box.

8

Firm the potting mixture around the plants then water well but gently. You may need to add more mixture as the contents of the box settle. Gently part the stems and sprinkle in small amounts by hand to avoid covering the blooms.

A TRANQUIL MINI-POOL

A miniature pool, created from a wooden barrel and surrounded
by cool green foliage, will add a refreshing extra dimension to
even the smallest patio or garden.

MATERIALS LIST

You will need: a wooden barrel (1); water plants such as Arrowhead, *Sagittaria sagittifolia* **(2) and Japanese Iris,** *Iris ensata* **(3); fine-mesh planting baskets (4); bricks (5); trowel (6); small brass nails (7); hammer (8); aquatic soil (9); gravel (10); floating plants such as Frogbit,** *Hydrocharis morsus-ranae* **(11) and Water Hyacinth,** *Eichhornia crassipes* **(12); algaecide (13); Monkey Musk,** *Mimulus luteus* **(14).**

Throughout history, water has played an important part in garden design. Whether splashing and vital, or still and tranquil, water adds a completely new dimension, with its capacity to reflect light.

Oriental art form

The ancient Egyptians cultivated Water Lily pools, and the Incas constructed water channels and basins from gold and silver. The water garden as an imitation of nature was perfected by the Chinese, and later the Japanese elevated it into an art form.

In European monasteries, gardens were often designed around the essential well, and a carp pool would provide food as well as an aid to contemplation. Formal water gardens flourished in Italy during the Renaissance, and in France the use of water as an elaborate element of complex garden design reached its peak in the 18th century. Romantic 'natural' English gardens at this time saw huge lakes being constructed and water features being carved into landscapes.

Unique qualities

Although few people can attempt ambitious water features today, even the smallest patio, city courtyard or suburban garden can still benefit from the unique qualities of water. A miniature pool can be created in any watertight container, to become home to a delightful collection of easily-grown water plants and a few Goldfish, which will add movement and interest.

Surrounding the pool with cool green foliage plants, such as Ferns and Bamboos, in attractive glazed pots will create a lush and inviting feature. Placed near garden seats, in a sunny position, even such a small amount of water will reflect the changing sky, providing a cooling atmosphere on hot Summer days.

Positioning

Plant up the barrel in its final position, as it will be impossible to move once it is filled with water. If a watertight barrel cannot be found, use a butyl rubber lining. An inexpensive alternative is heavy-duty black polythene, which would last a season, and could be replaced when the plants are repotted.

PICK & PLANT

Other plants for a small pool include:
- **Marsh Marigold** (*Caltha palustris*) with large yellow Buttercup-like flowers.
- **Pickerel Weed** (*Pontederia cordata*), which has spikes of blue flowers and heart-shaped leaves.
- Most Reeds and Rushes are too invasive to grow in a small space, but the dwarf form of **Reed-mace**, *Typha minima,* is attractive if kept under control.
- **Dwarf Water Lilies,** such as *Nymphaea pygmaea alba,* are a possibility, but are very vigorous growers.
- Some underwater plants, which help to oxygenate the water, also produce flowers above the surface; **Water Violet** has pretty mauve flowers and fern-like leaves.

CREATING THE WATER GARDEN

TIP

1

If using an old barrel, make sure it is watertight. Try to find a genuine barrel that has been used to hold liquid. Soak the outside thoroughly and fill with water so that the timbers swell, or preferably submerge it in water for at least 24 hours.

2

Most water plants are sold in fine-mesh planting baskets, but if any are growing in nursery pots, repot them into the special baskets, using aquatic soil. Cover the soil with a good layer of fine gravel to prevent it from washing out into the water (inset).

3

Use the Monkey Musk plants, with their Orchid-like yellow flowers, in a group at the back of the pool. Fix the planting baskets to the side of the barrel with small nails, so that the tops of the pots will be 5cm (2in) below the eventual water level.

4

Position the Japanese Iris in a group at the side of the pool; the pots can be raised to the correct level by propping them on top of plant pots or spare planting baskets. Again, ensure a depth of about 5cm (2in) of water above the pot.

COMPLETING THE WATER GARDEN

Complete the planting by placing one or two Arrowhead plants in a group at the other side of the pool. Bricks can be used to support the planting baskets at the correct level, or they can be nailed to the inside of the barrel as before.

Once all the water plants are in position, fill the barrel with water to a depth about 8cm (3in) below the rim. Keep a check on the amount of water used to make it easier to calculate the amount of algaecide which will need adding.

To prevent the growth of algae, which would quickly turn the water green, a specially-developed algaecide preparation should be used. Follow the manufacturer's instructions, ensuring that exactly the correct dosage is used.

Finally, plants such as Frogbit and Water Hyacinth, which simply float on the surface, should be placed in the water. Only one or two plants are needed, as they will spread rapidly. Allow the water to settle and clear before putting in Goldfish.

TUMBLING SWEET PEAS AND LOBELIAS

Fast-climbing Sweet Peas will add fragrance and a riot of colour to your patio. Team them with trailing Lobelia for a delightfully informal, swirling display

MATERIALS LIST
You will need: a half barrel, ready cleaned and prepared (1); crocks (2); Lobelia (3); 4 x 1.5m/5ft canes (4); Sweet Pea plants (5); potting compost (6); trowel (7); garden twine (8); string (9); horticultural grit (10).

The annual Sweet Pea (*Lathyrus odoratus*) is a fast-growing climber, which can be used for tub displays throughout the Summer: the fragrance of the flowers is exquisite and they last well as cut blooms. Some varieties of Sweet Pea grow to 2m (6ft 6in) tall and have to be supported by canes and twine. The plant produces tendrils with which it anchors itself to its support, but when the plants are small, they need assistance to reach the first anchor point, and can be loosely tied in place.

A helping hand
The number of plants used here will compete for space, and may need encouragement to stay in bounds; green garden twine is used to collect up all floppy growth, which is tied back to the main cane framework. This may appear scruffy, but within a week, the plants will have disguised the twine with foliage.

Blooms in abundance
The more Sweet Pea flowers you pick for use indoors, the greater number of blooms the plants will produce: like all annuals, this plant's aim is to set seed - as long as that process is delayed, it will continue to produce new blooms. All spent flower heads must be snipped off, and the plants may need checking at least every other day. A well-fed tub of plants like this will flower for about three months.

Most Sweet Pea plants offered for sale will be mixed colours and often with many plants in a pot. Carefully disentangle the plants from each other and plant immediately - before the minute root hairs begin to desiccate.

The plants should be pinched out a couple of times to encourage branching. Spray with a suitable fungicide for powdery mildew during hot, dry weather, and watch for slug damage.

Water-in the newly-planted Sweet Peas; as they grow, the need for moisture will steadily increase, and they may demand water twice daily in the heat of Summer. Do not allow the compost to dry out. Feed weekly with a general-purpose feed, beginning about a month after planting: the compost should be free-draining and of good quality.

Spoilt for choice
The Spencer range is a good choice: select from the red 'Garden Party' and 'Red Ensign', red-and-white striped 'Wiltshire Ripple', the wavy pink flowers of 'Pennine Floss', dual-coloured violet and deep blue 'North Shore', mid-blue 'Blue Danube', rich and creamy 'Lillie Langtry', and purest white 'Royal Wedding' and 'Snowdonia Park'. Each seed merchant has their own range; select for rich fragrance as well as colour.

PRACTICAL POINTERS

❑ **Prepare the barrel**
Rub-down rusty metal with a wire brush, then paint it; treat the wood with linseed oil and bore drain holes in the base (below).

❑ **Heavy work** Either fill the container in situ - placing it on pieces of tile to keep it clear of the paving - or be prepared to manoeuvre it into place using a sack barrow.

PREPARING THE TUB & CANES

1

To prevent the potting compost from leaching out through the drainage holes in the base of the barrel each time the plants are watered, spread a layer of crocks in the bottom. Make sure that the drilled holes are covered by the pot shards, but not blocked in any way.

2

Fill the barrel with moist compost (inset), 'rattling' the tub to encourage the compost to settle. Slightly over-fill the container to allow for further settlement when the compost is watered. Mix in some horticultural grit to improve the drainage qualities of the compost.

3

Push the ends of two of the canes into the compost at opposite sides of the barrel, then tie them securely together with stout string (about 10cm/4in from the top). Make sure that they are tightly bound. Add the remaining canes, spacing them evenly, and tie into place (inset).

4

Using the green twine, secure an end to a cane, about a third of the way up: encircle the canes, tying the twine to each one. Repeat the process two thirds of the way up the canes: the twine and the pyramid of canes form the plants' anchorage system, and must be secure.

PLANTING THE SWEET PEAS & LOBELIA

5

With the Sweet Pea plants still in their pots, arrange them symmetrically around the base of each cane, allowing three or four pots per cane. Lightly tap the side of each pot with the blade of a trowel to release the grip of the rootball and then carefully remove the plants.

6

Trowel out a hole big enough to take the rootball of each Sweet Pea plant: firm each plant gently into position, ensuring that it is placed only as deep as it was in the original pot. Lean larger plants against the twine and cane supports: small plants may need tying in place.

7

To provide a finishing touch to the half barrel, plant Lobelia (available in light blue, dark blue, red and white) around the rim so that it will trail over the edge and soften the lines of the container; pull each small clump of plants away from its neighbours and pop it into place.

8

Firm the compost well to the sides of the barrel, using your fingertips; add more compost if necessary. If you haven't made up the container in situ, move it into position before watering well: use a rose on the watering can to ensure that the plants are not flooded or dislodged.

A TOUCH OF THE TROPICS

Palms, despite their exotic appearance, survive well in a cooler climate, and when planted in classical terracotta pots can inject the patio with a flavour of the tropics.

PLANTING THE PALMS

You will need: A terracotta planter (1); a Cabbage Palm, *Cordyline australis* **(2); a trowel (3); potting compost (4); crocks (5); and a Fan Palm,** *Trachycarpus fortunei* **(6).**

Palms have always carried with them associations with tropical islands, exotic and exciting places, but some species can be grown perfectly happily outside in cooler climates. With their dramatic and delicate fronds they are ideal for a patio display which will last the whole year round.

PRACTICAL POINTER

❑ **Trimming the roots**
Garden centres usually keep Cordylines and Palms in pots sunk in compost on a sand layer, and the roots grow through the drainage hole. Trim off straggling roots with scissors before potting-on into the planter. This also makes removing the plant from its pot far easier, too.

1 Cover the drainage hole of the terracotta plant pots with crocks to aid drainage (inset). Then fill to about 2/3 the depth of the pot with a layer of ordinary free-draining container or potting compost.

2 Remove the Cabbage Palm from its pot and gently tease out the roots with your fingers to help them spread in its new container; place in the terracotta pot then fill with potting compost and firm down.

TIP

3 Repeat the process for the Fan Palm, making sure that the level of the compost is correct: if the plant is buried too deeply the crown may rot. Gently firm the compost around the rootball.

4 A windy site could result in browning of leaves at the base of the Palm (inset) and the delicate fronds of the Cordyline will also succumb to damage from the weather. Trim off dead growth with scissors.

ALL-WHITE DISPLAY

An elegant Victorian-inspired trough planted with a mass of cool white flowers and trailing silver foliage provides a refreshing contrast to Summer heat.

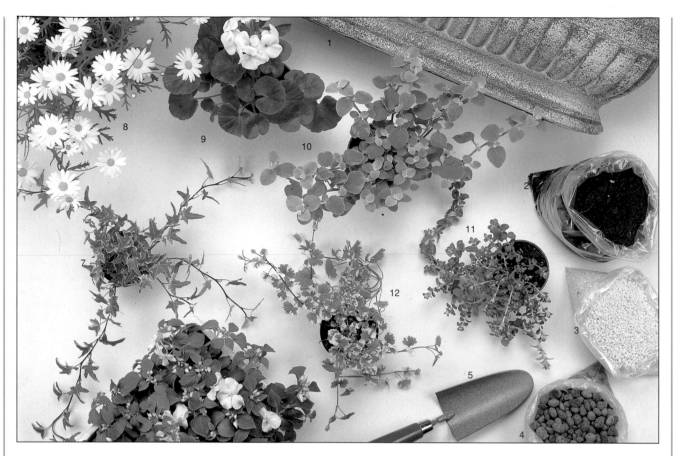

At the height of Summer, when the sunlight is strong and the garden bright with colour, an elegant trough standing on a terrace or patio, can provide a cool and refreshing focal point if planted with a restrained mixture of white and silver-grey plants. This subtle approach to annual planting can be particularly successful if the trough is positioned were it will be in partial shade; the light-toned flowers and foliage will then be seen to best advantage. White and silver are also colours which are very effective in the evening, when they shine out in the twilight, and can be appreciated to the full on a warm evening.

Authentic design

A planting trough for this delicate palette of colours should be chosen with care; most stone containers will be satisfactory, but if a plastic replica has to be the choice, either because of the cost, or weight if it is to be used on a balcony, then aim to choose a design and colour which will complement the planting. This reproduction of a Victorian planter is quite acceptable as it is based on an authentic design, and the grey colouring harmonizes with the foliage. (See Practical Pointer for tips on improving plastic containers.) This type of container planted in silvers and whites can look most effective in conjunction with white garden furniture, especially the many reproductions of Victorian cast-iron designs which are available today.

White and silver

Many popular and easily-grown plants are available in white flowering varieties, however they are often overwhelmed in gardens and nurseries by the mass of sometimes garishly-coloured, big-selling plants. Geraniums, Busy Lizzies, Fuchsias, Marguerites, Lobelias, Petunias, Begonias, Alyssum, Snapdragons and Tobacco Plants all have attractive white varieties, while silver-leaved *Senecio bicolour cineraria*, *Santolina* and *Artemesia* could be added to the list of possible subtle silver-foliage plants.

MATERIALS LIST
You will need: a long planter (1); potting compost (2); moisture-retentive granules (3); moisture-retentive clay pellets; (4) trowel (5); white Busy Lizzie, *Impatiens;* **(6); Variegated Ivy,** *Hedera helix* **(7);** *Chrysanthemum frutescens* **(8); white Geranium,** *Zonale Pelargonium* **'Century White' (9);** *Helichrysum petiolare* **(10);** *Helichrysum microphyllum* **(11);** *Nepeta hederacea* **'Variegata' (12).**

PRACTICAL POINTER

❏ **Distressed look** To improve the appearance of a plain white plastic container, try spraying it with medium grey matt paint, then immediately rub over with a cloth to remove some of the paint on the higher parts of the design. Experiment first so you master the technique, which can be very effective.

PREPARING THE PLANTING TROUGH

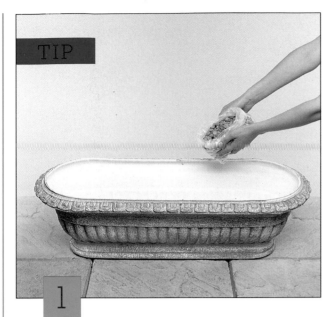

TIP

If the trough has drainage holes, place crocks over the holes first. To prevent excess water loss, place a 3-4cm layer of moisture-retentive clay pellets in the base of the trough; these will absorb water and release it into the potting compost as needed.

To retain additional moisture in the trough, mix a few handfuls of moisture-retentive granules into the compost, then put a deep layer into the trough. This will help lessen the weight of the trough, which could be an important factor on a balcony.

Begin the planting with the tallest plants, in this case the Marguerites, which should be positioned centrally, at the back of the trough, if it is going to be seen from one side only. Ensure that the plants are positioned at the correct level in the trough.

Next choose the longest trailing plants, in this case variegated Ivy, and plant one at each end of the trough, facing sideways, and another centrally, in front of the Marguerites, so that the Ivy trails down naturally over the edge of the trough.

PLANTING THE TROUGH

5 Plant white Geraniums beside the Marguerites, to form a contrast with the feathery grey foliage. Avoid a rigidly symmetrical arrangement of plants, but aim for a balanced distribution. If the trough is to be seen from all sides, check all round.

6 Use the two types of trailing Helichrysum along the front of the trough to eventually form a curtain of silver foliage. If it is to be viewed from all sides, plant more at the back of the trough. Fill in gaps in the compost as the planting proceeds.

7 Any gaps around the rim of the trough can now be filled with variegated *Nepeta*, grouping plants together to form effective areas of delicate foliage which will cascade over the planter as the Summer goes on. Fill any gaps in the compost.

8 Complete the planting with a few white Busy Lizzies (*Impatiens*) to fill in any obvious gaps. Fill all the remaining spaces in the compost and carefully firm down, taking care not to damage any of the plants, and water the trough thoroughly.

ORIENTAL DELICACY

The delicate purple foliage of the Japanese Maple complements
perfectly the Oriental styling of this glazed planter to produce a
display of subtle colour.

MATERIAL LIST
You will need; a large Oriental-style planter (1); lime-free compost (2); crocks (3); Japanese Maple, *Acer palmatum* 'Atropurpureum' (4); trowel (5).

The dark and spidery foliage of the *Acer palmatum* 'Atro-purpureum' makes an excellent choice for a sheltered spot in a town garden or patio. When containerized it will grow to no more than a manageable 2m (6ft) although the plant can naturally grow to around 6m (20ft).

Acer palmatum likes a moist, well-drained, lime-free soil. Position the pot away from drying cold winds as too exposed a position could result in leaf scorch.

The dark glaze and Oriental styling of the planter combine perfectly with the Summer foliage of this plant, which becomes progressively deeper in colour as the Autumn draws on.

PRACTICAL POINTER

❏ **Any dead branches will spoil the appearance of the plant, so remove them with secateurs.**

1

Place the container in its final position as, once it has been filled with soil and the Acer planted, it will be more difficult to move. Prepare the pot by lining the base with a layer of crocks, which helps to aid drainage.

2

Fill the planter with compost to within about 15cm (6in) of the rim. *Acer palmatum* 'Atropurpureum' is quite happy in most soils, even heavy clays; however, it does need a free-draining, lime-free growing medium.

TIP

3

Water the Acer a few hours before you transplant it . Before removing it from its plastic pot, you should pull out any weeds from the surface of the compost, so that they are not transplanted with the Acer.

4

Using the trowel, make an indent in the surface of the compost large enough to accept the root ball.. Fill around the plant with compost, slightly over-filling to allow for settlement. Compact gently so the plant is held securely.

SPRING BLOOMS

A late Spring display of Broom, Marguerites and Pansies will provide colour all Summer. It is ideal for a town garden where the tender Marguerites will not be caught by late frosts.

MATERIALS LIST
You will need:
Pansies, *Viola* **x**
wittrockiana
hybrids (1);
compost (2);
trowel (3); crocks
(4); terracotta
trough (5); Broom,
Cytisus **x** *praecox*
(6); tender
Marguerite,
Argyranthemum
foeniculaceum **(7).**

I t is often tempting, when wandering around a nursery or garden centre, to create a spontaneous containerized display from the large range of flowering plants in season at that particular time. Usually it is possible to find many combinations of shrubs and smaller flowering plants that look really stunning together.

An effective combination

One example of this type of display is the combination of Broom, Marguerites and Pansies shown here: the mass of buttery-yellow Broom flowers makes an ideal backdrop to the brightly coloured Pansies, while the dainty Daisy flowers and fine foliage of the Marguerites serve to soften the whole effect.

A profusion of flowers

Although the Brooms will have only one flush of flowers, they will more than earn their keep by the shear number of flowers they produce. With regular dead-heading, the Pansies and Marguerites will flower all Summer and will be offset nicely by the rich green Broom foliage.

The plants will flower most profusely if placed in a bright, sunny position. The compost should be kept moist, but not soggy. Feed regularly during the Summer months, using a proprietary fertilizer, to get the best from the plants.

PRACTICAL POINTER

❏ **Protection from frost**
Although frost is an unlikely occurrence in late Spring, place the container initially in a sheltered position, until all risk of frost has passed, as the Marguerites used here are not frost-hardy. This precaution will not be necessary in a town garden, due to the protective nature of buildings, which cause a beneficial microclimate.

PREPARING THE TROUGH

1 Place a good handful of pebbles or pieces of broken flowerpot into the bottom of the trough. Make sure that you cover all the drainage holes to prevent the compost from leaching out through the holes when the display is watered.

2 Use a free-draining compost in the trough, adding up to a quarter by volume of horticultural grit to the mixture, if necessary, to improve the texture. Fill the trough to within 13cm (5in) of the rim and firm lightly with your hands.

3 Remove the Broom plants from their pots by turning upside down and tapping each pot sharply to release the rootball. If the roots of the plants are growing round in circles, tease a few of them out straight before planting in the trough.

4 Lower the roots of the first Broom plant into the trough. Check that the top of the rootball is level with the rim of the trough; if it is not, adjust the amount of compost underneath. Plant one Broom plant at each end of the terracotta trough.

PLANTING THE TROUGH

5

Remove the Marguerites from their pots and use them to fill the spaces between the Broom plants along the back of the trough, leaving the space right at the front for the Pansies. Top up the level of the compost first, if necessary.

6

If the Pansies have been grown in a polystyrene strip, remove them gently by pushing a finger or thumb through the polystyrene under each compartment. This will release the plant, but take care not to damage it in any way.

7

Plant the Pansies in the space left in the front of the tub, and in any other spaces left between the larger plants. Position them close together for an immediate effect, making sure that the compost is no longer visible and the trough is full.

8

Firm the compost in the trough by pushing fingers down between the plants, making sure that there are no air spaces, and filling any gaps you come across with more compost. Water the plants in thoroughly to help them become established.

TEMPTING TOMATOES

Tomatoes are an ideal crop to grow in a container on the patio.
There are many varieties to choose from, providing a delicious
alternative to commercially-grown produce.

MATERIALS LIST
You will need: a large flower pot (1); small flower pots (2); cork or small saucer for firming (3); seed compost (4); pencil (5); cling film (6); rule (7); tomato seeds (8); name tags (9); sieve (10); timber (11); grown-on tomato seedlings (12); cane holders (13); trowel (14); 64 galvanized nails - 4cm/1.5in (15); 22 galvanized nails - 5cm/2in (16); 8 screws - 5cm/2in (17); electric drill (18); paint brush (19); screw driver (20); sharp knife (21); hammer (22); 3 canes - 156cm/5ft (23); gro-bag (24); wood preservative (25).

Tomatoes are a lovely salad crop to grow at home - the fruits are always sweeter and juicier than those available from shops, as you can let them ripen naturally on the plant in the warm sunshine.

Many different types of tomatoes are currently available from seed companies, ranging from the bite-sized, deliciously-sweet Cherry varieties to the huge, meaty Beefsteak tomatoes, and from the traditional red varieties to golden yellow ones and even tomatoes with stripes.

An easy option
To make home-growing even easier, tomato plants can be raised on the patio in gro-bags, which are convenient, mess-free, and ideal for someone without a lot of space. Gro-bags, however, can be unattractive, but a simple box-shaped container, such as the one shown here, is easily made and can be used for many seasons to come.

Sow the seeds in mid-April, keeping them warm (18°C/65°F) and dark until germination occurs. As soon as the seedlings appear, move them to a bright place, such as a sunny windowsill.

At the end of May, when all risk of frost has passed and the plants are 15-22cm (6-9in) tall, transplant them into the gro-bag outside. Place the bag in its box, in a sheltered, sunny spot.

Water with care
Gro-bags contain a light peat-based mixture that is difficult to keep moist without being too wet. Water sparingly and often - drying out causes flower-drop, waterlogging kills the roots, and irregular watering will cause the fruits to split as they ripen.

Feed is also important: gro-bags contain enough nutrients to support the plants until they produce their first truss of tomatoes. After this, feed with a proprietary tomato fertilizer.

A fruitful harvest
Tomato plants growing outdoors probably won't produce more than about five trusses of fruit in a

MAKING THE GRO-BAG BOX

Simple container The wooden container for the gro-bag is made in a simple box shape. Four upright posts form the corners; two bearers run lengthways between the uprights on each side and support the base slats. The box is completed by slats that are nailed between the posts on all four sides.

❑ **Materials list** The following timber is required:
2 bearers 5 x 2.5 x 108cm (2 x 1 x 43in).
4 uprights 5 x 5 x 53cm (2 x 2 x 20in).
11 base slats 8 x 2.5 x 30cm (3 x 1 x 12in).
8 end slats 8 x 1.25 x 41cm (3 x 1/2 x 16in).
8 side slats 8 x 1.25 x 108cm (3 x 1/2 x 43in).

season. To ensure the healthy development of these trusses, pick off the plant's growing tip, two leaves beyond the fifth truss when it has formed, and regularly pick out any side shoots that grow.

The fruit should be ready for eating from early July onwards. To pick, hold each fruit in your palm, and with your thumb, press the slight swelling on the stalk just above the fruit - it should break off cleanly.

Before the first frosts arrive, usually in mid-September, remove any trusses that have not yet turned red and lay them indoors on a sunny windowsill where they will continue to ripen.

PICK & PLANT

Tomatoes are generally sensitive to cold, but many varieties can be grown outside from late May to September in a sunny, sheltered position.
Cherry varieties:
❏ **Sweet 100** - produces long trusses of very sweet fruits which mature early.
❏ **Gardener's Delight** - one of the finest flavours available and a heavy-cropping variety.
❏ **Sungold** - produces unusual golden-orange, bite-sized fruits.
Medium-sized fruits:
❏ **Money Maker** - extremely reliable; produces a heavy crop.
❏ **Marmande** - large, irregular, fleshy fruits having few seeds and a distinctive flavour.
❏ **Golden Sunrise** - bright yellow fruits with a sweet and fruity flavour.
❏ **Tigerell**a - orange and red striped fruits with a good, tangy flavour.
Large fruits:
❏ **Jumbo Tom** - a beefsteak tomato with a delicious flavour and a heavy yield.

SOWING THE SEEDS

1

Sieve some seed compost onto a plastic sheet and discard the pieces left in the sieve (inset). Use the sieved compost to fill a 15cm (6in) pot. Fill the pot to the top and tap it firmly on a hard surface to settle the compost.

2

Strike off the excess compost with a rule (inset). Using a saucer or similar flat object, such as a large cork, firm the compost gently and evenly to form a flat and slightly compacted surface on which to sow the seeds.

3

Carefully tear a small strip from one end of the seed sachet. Hold the sachet in one hand and tap it with the other hand to shake the seeds out onto the compost. Sow only about 20 of the seeds, well spaced, in the pot.

4

Put some of the sieved compost back in to the sieve and tap it gently over the seeds to form an even layer about 1cm (1/2in) deep. Make sure it covers all the seeds to the same depth. There is no need to firm this layer.

PLANTING THE SEEDLINGS

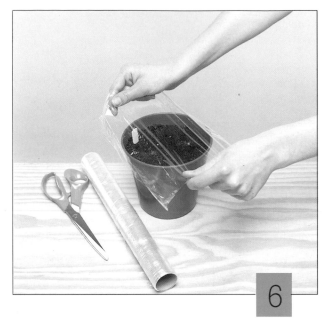

5 Water the seeds thoroughly, using a watering can with a fine rose, taking care not to dislodge the compost and the seeds underneath with heavy droplets. Continue until water starts to drip from the bottom of the pot. Using the name tag, label the seeds and note the sowing date (inset).

6 Cut off a piece of cling film and use it to cover the seed pot to keep the moisture in and prevent the need for watering again until after germination. Put the pot in a warm, dark place until the seeds germinate in about 7-10 days, then move them to a bright position to grow on.

7 Although you will need only three plants, it is best to grow extra in case some are unsuccessful. Fill five 8cm/3in pots with sieved compost (inset). Insert a pencil beneath the roots of each seedling, hold the plant by its leaves and gently lever it from the pot, trying not to disturb the roots.

8 Make a hole in the centre of the compost in each pot, using the end of the pencil and lower the tomato seedling roots into the hole. Make sure the seedlings are planted at the same depth that they were in the original pot. Lightly push compost around the roots with the pencil.

BUILDING THE CONTAINER

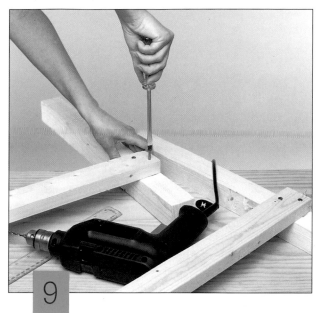

Begin to assemble the box by making the frame, screwing a bearer between each pair of uprights. Make sure the bottom edge of the bearer is 8cm (3in) from the end of each upright. Drill two holes at each joint through which to drive the 5cm (2in) screws.

Join the side frames together with the base slats, nailing the ends of the slats to the tops of the bearers with 5cm (2in) nails. Fix the two end slats first, with the outside edge of each flush with the ends of the bearers and uprights. Space the others 2cm (1in) apart.

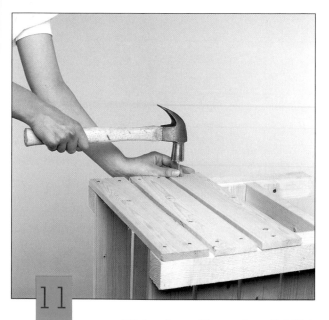

Fit the slats with two 4cm (1 1/2in) nails at each end. Fit the side slats to the insides of the uprights, with the upper edge of the top one flush with the tops of the uprights. Leave 1.25cm (1/2in) gaps between slats. Fix the end slats to the outsides of the uprights, spaced in the same way.

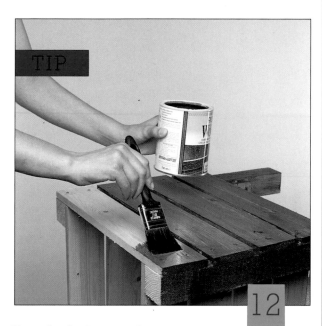

TIP

To make the box weatherproof, treat it with a wood preservative. Wood preservatives are available in many colours, including green, blue and many shades of brown - choose a colour that will complement your patio. Apply it according to the maker's instructions.

PLANTING THE TOMATOES

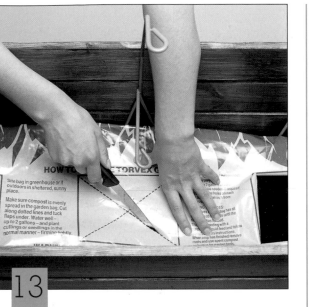

13

When the preservative is completely dry, you can put in the gro-bag and plant it up. Lay the bag in the box with the side which has the holes marked on it uppermost. Position the cane supports and, using a sharp knife, cut out the holes for the plants in the positions marked.

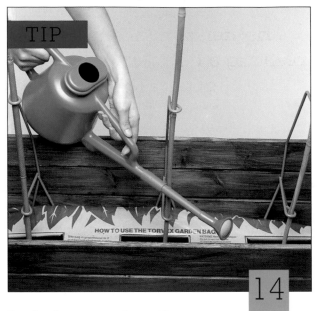

TIP

14

Put the three canes in position, threading them through the loops of the cane supports and pushing them into the compost. Before planting the tomato plants in the gro-bag, thoroughly moisten the compost, as it may be difficult to wet sufficiently with the plants in place.

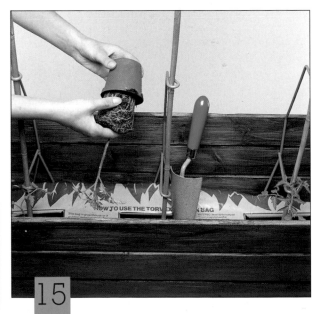

15

Choose the three sturdiest tomato plants and remove them from their pots, causing as little root disturbance as possible. Plant them in the three holes in the gro-bag. Water the plants thoroughly, but avoid making the compost soggy. Discard the remaining plants.

16

As the plants grow, side shoots will develop in the junctions between leaves and stem. Remove these as soon as possible, otherwise they will take nutrients from the fruit-bearing parts of the plant. Tie the plants loosely to the supporting canes using garden twine.

SUNNY CLIMBER

Brighten a plain exterior wall during the Summer months by planting a hanging basket with the lush foliage and vivid blooms of the aptly-named climber Black-Eyed Susan.

PLANTING THE DISPLAY

MATERIALS LIST

You will need: Black-Eyed Susan, *Thunbergia alata* **(1); hanging basket (2); moisture-retentive clay pellets (3); compost (4); trowel (5).**

Coming originally from tropical Africa, this vibrant climber is excellent to cheer up a plain wall during the Summer months. Black-Eyed Susan is a very fast grower and produces a profusion of blooms from early Summer right through to the Autumn. The tubular flowers are 5cm (2in) across and are white, yellow or shocking orange in colour, with a distinctive dark chocolate-coloured eye.

For best results, hang the basket in a very sunny position, providing light shade on really hot days. Water freely, ensuring that the compost is moist at all times, but not soggy. Hang the basket outside only when the danger of frost has passed - about mid-May in the UK.

In the Autumn, after flowering has finished, discard the plant and start again with a new one in the Spring.

PRACTICAL POINTERS

❏ **Pick out the tips** of young plants to promote branching and a fuller, more bushy shape.
❏ **Remove faded flowers** before the seed is produced in order to encourage more flowers to appear.

1

With the basket hook and supports removed, put a 3cm (1in) layer of moisture-retentive clay pellets into the bottom of the basket to aid drainage (inset). Trowel in compost until the basket is about two-thirds full.

2

Remove the plant carefully from its pot by putting your fingers on each side of the plant base then, with the other hand, turning the pot upside down and tapping it on a firm surface, such as a tabletop.

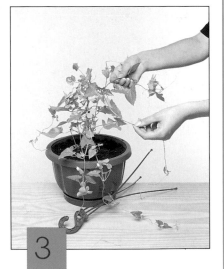

3

Ease the rootball into the basket, firming the compost around the roots and adding more to fill to just below the rim. Tease the plant shoots apart, separating them as much as possible. Remove any supporting sticks.

4

Fix the support back on to the basket and wind the shoots up around it, taking care not to damage flowerbuds and new side shoots. Arrange the shoots evenly around the basket for an overall spread of flowers.

BLUE SUMMER HAZE

When the Summer bedding is starting to fade, fill a barrel with an elegant mixture of feathery foliage and blooms in a haze of blues and silver that will shine in the late Summer sun.

After the Summer blooms have finished, and before the rich tones of Autumn arrive, there is generally little colour outside to cheer the gradually fading days. It is possible, however, to find a whole host of plants which look their best at this time of year.

Shown here is the beautiful Caryopteris (*Caryopteris x clandonensis*), an elegant shrub which produces a profusion of deep lilac-blue blooms in late Summer. It is teamed with clear vivid blue Michaelmas Daisies and Violas. The strong colour is complemented by silver-leaved plants - Cotton Lavender and Artemisia - which have fine feathery foliage that is evergreen and will make a good show through the Winter months.

Caryopteris is a deciduous shrub, and not very attractive in the Winter. After it has finished flowering, it can be replaced in the tub with a blue conifer, which will maintain its colour all year. When the Violas finish, simply replace them with Winter-flowering varieties.

Place the barrel in full sun to encourage flowering, and where the silver foliage will shine. Dead-head the Violas regularly to encourage them to produce more blooms.

MATERIALS LIST
You will need: a half barrel (1); Cotton Lavender, *Santolina chamaecyparissus* (2); Michaelmas Daisy, *Aster* 'Lady in Blue' (3); blue bedding Viola (4); crocks (5); trowel (6); compost (7); *Artemisia arborescens* 'Powis Castle' (8); *Caryopteris x clandonensis* (9).

PICK & PLANT

Other suitable blue-flowered plants which bloom into the late Summer and Autumn include:
- ❏ **Blue Marguerite** (*Felicia amelloides*) - a bushy, low-growing shrub that produces a profusion of bright blue, Daisy-like flowers, all Summer and into the Autumn.
- ❏ **Hebe** - these handsome, glossy-leaved shrubs have the advantage of being evergreen; cultivars such as *H.* 'Autumn Glory' and *H.* 'Purple Queen' are a suitable size for a tub and produce purple-blue flowers in late Summer and Autumn.
- ❏ **Californian Lilac** - *Ceanothus* 'Burkwoodii' and *C.* 'Gloire de Versailles' are both species with deep green leaves and clear, pale blue blooms; a very attractive plant.
- ❏ **Swan River Daisy** (*Brachycome iberidifolia*) - this bushy, Autumn-blooming annual has fragrant, Daisy flowers; many forms are available.
- ❏ **Violas and Pansies** - many blue Violas and Pansies are available in flower from garden centres and nurseries during the late Summer months.

Suitable evergreen, silver-foliage plants include:
- ❏ **Common Rue** (*Ruta graveolens* 'Jackman's Blue') - with bushy, finely-divided aromatic foliage.
- ❏ *Hebe pinguifolia* **'Pagei'** - has small, rounded, stiff leaves on woody stems and forms a silver cushion.
- ❏ **Dusty Miller** (*Senecio maritima*) - usually grown as an annual for Summer bedding, this deep-lobed bushy plant is actually evergreen and useful for Winter colour in mild areas.
- ❏ **Curry Plant** (*Helichrysum italicum*) - with linear, aromatic grey foliage, this handsome shrub is a popular choice.

PREPARING THE BARREL

1 Begin by checking that the barrel has drainage holes. If it has not, drill three holes using a 3cm (1 1/4in) wood bit, in a line at the bottom of the barrel, taking care not to split the wood. Put a handful of large crocks over each hole to prevent compost leaching out during watering.

2 Fill the barrel with compost, up to 5cm (2in) below the rim, firming gently as you go to form a level surface. Use a free-draining compost, adding up to a third by volume of horticultural sand or grit to the mixture, if necessary, to improve the texture and drainage qualities.

TIP

3 Arrange the plants, still in their pots, in the barrel to decide on planting positions. Here, two Caryopteris and two Artemisia form the back row, with two Cotton Lavender plants and two Michaelmas Daisies in the middle row, and the Violas filling the remaining area in the front.

4 Position the back row of plants first. Remove the Caryopteris from their pots and gently tease out the roots a little so they no longer form a pot shape; this will encourage outward root growth in the barrel. Plant them together with the tops of the rootballs level with the barrel rim.

COMPLETING THE BARREL

5

Firm compost gently around the rootballs of the two Caryopteris plants and add more, if necessary, to bring up the compost level around them. Remove the two Artemisia plants from their pots by turning them upside down and tapping the pots firmly to release the rootballs.

6

Position one Artemisia plant on each side of the two Caryopteris, close to the sides of the barrel so that the foliage will spill out and soften the hard edges. Plant the rootballs at the same level as those of the Caryopteris, and add more compost between them to form a level surface.

7

Moving on to the middle row, plant the low-growing Cotton Lavender plants in front of the Caryopteris so that the blue flowers are not hidden. Position the slightly taller Michaelmas Daisies on each side where the Artemisia plants will form a silvery backdrop to the Daisy flowers.

8

Remove the Viola plants carefully from the strip, taking care not to damage the roots. Plant them evenly in the remaining space in the front of the barrel, bringing some right up to the front rim to soften it a little. Water the plants in well using a watering can with a fine rose attached.

STRAWBERRY FARE

Ideal for a patio, balcony or even a large windowledge, this compact, multi-pot container is an attractive and labour-saving means of growing Strawberries.

**MATERIALS LIST
You will need:
trowel (1); self-assembly
Strawberry
towerpot (2); peat-based potting
compost (3); grit
(4); Strawberry
plants (5).**

Strawberries grown at home always seem to taste far better than any you may buy, and this self-assembly towerpot of four stacking planters will enable you to grow up to 13 Strawberry plants in a limited space, as it is a mere 59cm (23in) tall. There is no messy straw to tuck under the fruits, as most of the Strawberries will cascade over the edges of the planters.

The unit can be planted at any time of the year, but for a small crop in Summer, the Strawberries should be in position by late Spring. They are sold as bare root plants in Winter and early Spring, and in containers in late Spring and Summer. 'Royal Sovereign', 'Cambridge Favourite' and 'Gorella' are old favourites - grow a mixed batch for a range of flavour and cropping times; always buy certified virus-free stock.

Integral reservoirs

Because the towerpot has an inbuilt reservoir at the bottom of each layer and a capillary wick, watering is unlikely to be a problem. This allows the compost to remain moist without saturation. During periods of prolonged rainfall, it may be necessary to tilt the tower to prevent flooding.

Sunny spells and fresh air

Place the towerpot where there is a fair degree of direct sun. Strawberries are reluctant to flower and ripen unless in direct sun for at least part of the day: rotate the pot every fortnight to ensure that all the plants receive an even amount of sunlight.

Wet leaves and generally muggy conditions will encourage Botrytis. If possible, keep water off the leaves and fruits, and ensure that the plants are getting some fresh air.

Feed and care

During the first season in fresh compost, the plants will need no extra feed to make leaf growth; in the following two years, feed them with a general-purpose plant food every two weeks from mid Spring to early Autumn. From the first Summer, also feed all the plants every other week with a high-potash feed, from mid May until the end of fruiting.

The fruiting period usually ends by August, at which time the plants should be trimmed to remove all the old leaves, stems and so on. The plants will make a lot of new growth before the onset of Winter, and will look much better for a 'haircut'.

Pests and diseases

Strawberries are prone to several problems: as they are a food crop, select any insecticide/fungicide with care. Aphids will attack the succulent ends of the runners, while hot, dry weather will encourage Powdery Mildew; warm, damp conditions are ideal for Downy Mildew and Botrytis.

PRACTICAL POINTER

❏ **Take care** If properly cared for, your Strawberry tower will provide you with a crop for three years: during the first season, only allow the plants to produce a few berries and trim off any over-vigorous runners - the whippy shoots which grow from the parent plant and are its method of regeneration.

The following two seasons the plants should crop well, and you should plan for replacement plants in the third season by pegging down a number of the runners into pots of compost - hold the baby plants in place with a hoop of wire. Once rooted, sever the baby from its parent and grow it on. If there is any sign of virus, discard all the old plants and buy fresh stock.

BUILDING THE TOWERPOT

Read through the assembly instructions and identify all the elements of the towerpot before beginning. Insert both ends of the wick into the slots in the platform, ensuring that they project evenly below the platform.

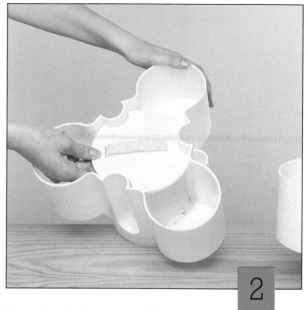

Drop the platform and wick assembly into place in one of the planters, making sure that its rim faces downwards; then complete the assembly of the lowest planter by pushing it into position on the supplied base.

TIP

To help improve the drainage properties of the potting compost, mix in up to a third by volume of horticultural grit; add some water at this stage if the compost seems dry. A plastic washing-up bowl is ideal for this purpose.

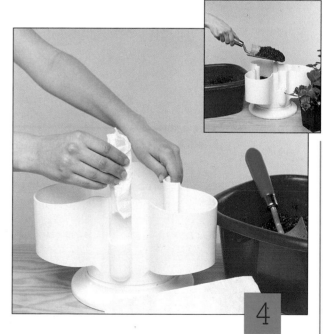

Before filling the planters with the compost mix, the float housings should be plugged with soft paper to ensure that they don't become blocked with compost. Trowel some compost into each planting bay, half filling the bay (inset).

PLANTING THE STRAWBERRIES

5 Set the plants in place, tucking them on the outer edge, tipped away from the centre. Add more compost (inset) and firm the plants and compost with your fingers: compost in the centre of the planter should be about 2cm (1in) lower than at the rim.

6 Repeat the process for the other planters and slot one on top of the other, using the integral locking slots: ensure that the float housings are on the same side of the tower for ease of watering. The top layer takes the 13th plant in the centre.

7 The floats act as water-level indicators for each planter's reservoir. Unplug all the float housings: assemble each float and its retainer, and insert into the float housings, placing the retainers so that their slots face down.

8 Carefully fill the reservoirs via the float housings. Once they have been filled, gently water-in the plants from above, using a watering can fitted with a rose. Do this gradually over a couple of hours, as the compost take-up of moisture can be variable.

EVERGREEN SHRUB TUB

A handsome terracotta planter filled with hardy evergreen flowering shrubs will bring a surprising amount of colour and texture to an otherwise barren Winter garden.

MATERIALS LIST
Large, frost-proof terracotta tub (1);
crocks (2); potting compost (3);
Hebe ellipticca variegata **(4);**
Skimmia japonica rubella **(5);**
Euonymus ovatus Aureus **(6);**
Euonymus japonicus microphyllus
variegatus **(7);** *Cineraria maritima*
'Silver Dust' **(8); Ivy,** *Hedera helix* **(9).**

When Winter robs the garden of its interest, then hardy, evergreen flowering shrubs come into their own. These tough, tolerant plants create abundant colour, texture and shape.

The variegated varieties of *Euonymus* and the shrubby *Veronicas* or *Hebes* are all particularly attractive - their interesting yellow, white or golden-edged leathery leaves will glow on the darkest days, while *Skimmia japonica rubella* is especially valuable for its red winter buds and white flowers in Spring.

Any of the many trailing types of Ivy can be used at the front of the group, and the whole display will benefit from the addition of the silvery foliage of *Cineraria maritima* as a highlight.

PRACTICAL POINTER

❑ **A compact shape** To maintain a compact grouping, cut the shrubs back well in the Spring, while preserving the natural shape of the individual plants. Feed with a liquid fertilizer.

1 It is important to choose a terracotta pot that is guaranteed frost-proof. Put a layer of crocks over the drainage hole and fill with potting compost to about two-thirds full (inset).

2 Start to plant from the back, arranging the *Euonymus ovatus* and *Skimmia* at each side of the central *Hebe*. Dig holes for the root-balls, insert the plants, and firm the compost around them.

TIP

3 Place the 'Silver Dust' in the centre of the planter, and firm down the compost. Plant a pair of the smaller, squatter *Euonymus japonicus* in the fore-ground. Back-fill with compost as you go.

4 Finally, plant the Ivy at the front, and fill-in any last gaps with compost. Firm-down carefully and water-in well. The shrubs should now form a dense group of nicely contrasting foliage.

TERRACOTTA TIERS

By slipping a few different-sized pots one inside the other, you can form a tower of colourful flowers and foliage that is ingenious and inexpensive to create.

MATERIALS LIST
You will need: three different-sized pots (1); selection of bedding plants, such as Osteospermum (2), Ivy (3), Brachycome (4) and Verbena (5); crocks (6); compost (7); trowel (8).

PRACTICAL POINTERS

❏ **The right dimensions**
Choosing the right pots to create the tower is all important. Make sure you have at least three different-sized pots, and check that each pot will fit inside the next largest pot with quite a lot of room to spare, providing enough space to plant a good range of bedding plants. Leave a gap of at least 10cm (4in) on all sides for the plants.

❏ **As high as you like**
You can build the tower as high as you like, but remember that the smallest pot must be large enough to contain at least one plant comfortably. The larger the bottom pot, the taller the finished tower can be.

Ready-made tower pots and herb pots with holes in the sides are expensive, and it is often difficult to find the size you are looking for, as choice is limited. But you can make something similar quite easily.

Ingenious idea
By slipping a few plain flower pots, one inside the other, it couldn't be easier to build your own tower pot, creating a sizeable feature at relatively little expense. Each pot is sunk half way into a slightly larger pot, so forming a tower. The plants are planted in the space around each pot, creating a fountain of colour.

Ideal for summer bedding
A tower pot lends itself well for use with Summer bedding plants, especially the trailing varieties which soften the hard edges of the pots.

Choose from Trailing Geraniums, Brachycomes, Ivies, the many forms of Verbena, the well-known Lobelia and sunny Nasturtiums for the sides; using a bushier plant at the top of the tower, such as an Osteospermum, Gazania or a group of dainty Busy Lizzies.

A question of watering
Careful watering is important with this design, especially if the pots you use are terracotta, as moisture will soak through each pot from the compost in the pot below. Before each watering, sink your fingers into the compost in the bottom pot for an indication of how much water to give at the top; adjust the amount accordingly.

PREPARING THE TOWER

1 Place a good handful of large crocks, such as pieces of broken clay flower pot or tiles, into the bottom of the largest pot, making sure that you cover the drainage hole to prevent compost from being washed out. Repeat for the other pots.

2 If you are planting Summer annuals in the pots, use a free-draining compost, if necessary adding up to a third by volume of horticultural grit. Fill the largest pot to about half-way with the compost and firm down thoroughly.

3 Take the medium-sized pot and stand it on the compost inside the large pot. Add or remove compost until it protrudes about 15cm (6in) above the rim of the larger pot. Check that the medium pot is standing level and not leaning at all.

4 Plant a selection of trailing plants, such as Ivy and Verbena, in the large pot in the space around the edges of the medium pot, where they will trail down and soften the edges. Fill the remaining space between them with more compost and firm lightly.

COMPLETING THE TOWER

5

Fill the medium-sized pot to about half-way with compost, firm thoroughly, then place the small pot inside it, positioning as before. Plant up the available space in the medium pot with more trailing plants, such as the Brachycomes.

TIP

6

Before planting the topmost pot, check that all the pots are held securely in place by the compost in the pot below. If any of the pots can be wobbled, pack more compost around it, taking care not to damage the roots of the plants.

7

Half-fill the top pot with compost and firm it lightly. Remove the plant you have chosen for the top position from its original pot, taking care not to damage its roots. Lower it into position, adding more compost and firming around it gently.

8

Water the plants thoroughly, making sure that all the compost in all of the pots is moistened. Use a watering can fitted with a fine rose to prevent compost from being washed over the edges of the pots by a heavy stream of water.

WOODLAND WONDERS

The graceful blooms of the Hellebore come in a beautiful range of sumptuous, dusky shades and are a welcome sight when they appear in the depths of Winter.

MATERIALS LIST
**You will need: wooden trough (1);
bark chippings (2); compost (3);
selection of Hellebores,** *Helleborus*
**species (4); broken flower pot (5);
hammer (6); trowel (7).**

Hellebores are probably the most sumptuous of Winter- and early Spring-flowering perennials. Their sculptural blooms appear in shades from the purest white, through pale green to dusky pink, purple and an almost navy blue, often with speckling inside.

The rich, deeply-divided foliage is usually evergreen, and several different species planted together will produce a handsome show throughout the year.

These woodland plants will be displayed to best advantage in a rustic trough, preferably in semi-shade. Keep the compost moist, but not soggy.

PICK & PLANT

Several species of Hellebore are available:
❏ **Stinking Hellebore** (*H. foetidus*) - clusters of pale green, cup-shaped flowers with red margins.
❏ **Christmas Rose** (*H. niger*) - large, flat, white blooms which often turn pink with age.
❏ **Lenten Rose** (*H. orientalis*) - flowers in shades of white, pink and purple, often with speckles inside.
❏ **Green Hellebore** (*H. viridis*) - dark green leaves and glowing, bright green blooms.

1

Lay a few pieces of broken flower pot on a firm surface and break them up with a small hammer until the pieces are roughly 5cm (2in) square. Place the pieces in the bottom of the tub, particularly over the drainage holes.

2

Hellebores will thrive in a humus-rich compost: add up to a quarter by volume of leaf mould, or manure, to a free-draining compost and combine thoroughly. Fill the trough to within 15cm (6in) of the rim, firming as you go.

3

Remove the plants from their pots and arrange on top of the compost, with the taller varieties, such as the Stinking Hellebore, at the back. When you are satisfied, fill in the spaces with more compost and firm all round (inset).

4

A bark mulch on top of the compost will not only look attractive, but also will prevent the compost from drying out too quickly in warm weather. Sprinkle on a layer of chips, about 2.5cm (1in) deep, taking care not to bury the plant crowns.

PURPLE AND YELLOW DISPLAY

A daring combination of purples and yellows will bring a vivid splash of colour to your patio that will last all Summer long, filling the air with the sweet scent of cherries.

MATERIALS LIST

You will need: an ornate urn (1); Cherry Pie plants, *Heliotropium arborescens* **(2); crocks (3); compost (4); Busy Lizzies,** *Impatiens walleriana* **(5); trowel (6); Slipper Flowers,** *Calceolaria integrifolia* **(7).**

Deep mauve Cherry Pie plants, purple Busy Lizzies and vibrant yellow Slipper Flowers, in a decorative urn, are an unusual combination. However, instead of clashing, the colours seem to complement each other, producing a stunning display to brighten up a patio.

The Cherry Pie plant is so named because the beautiful fragrance of the flowers is reminiscent of cooked cherries. The plant is usually grown as an annual, so along with the Busy Lizzies and Slipper Flowers, it should be discarded at the end of the season, after you've taken cuttings to root for next year.

Place the urn in a bright, sunny position, out of strong winds. Water freely and feed weekly while the plants are flowering. Encourage bushy growth by pinching out the growing tips of the Cherry Pie plants and dead-head regularly to produce fresh blooms.

PRACTICAL POINTER

❏ **Once full** of compost and plants, the urn will be extremely heavy, so if possible, plant it up in its chosen final position.

PLANTING THE DISPLAY

1

Place a handful of large crocks into the bottom of the urn to cover the drainage hole and prevent compost from leaching out during watering (inset). Add compost up to about 3cm (1in) below the rim, firming as you go.

2

Remove the Cherry Pie plants from their pots by tapping each pot sharply with the trowel to loosen the rootball. Plant the two plants close together in the centre of the urn to leave maximum space for the other plants.

3

Remove the Slipper Flowers from their pots in the same way and plant at equal intervals around the edge of the urn. Make sure the compost in the urn comes no higher up the plant than the compost in its original pot.

4

Fill in between the Slipper Flowers with Busy Lizzies, planting them close together for an immediate effect, filling the urn and covering the compost completely. Firm the compost between the plants to secure them.

AN ANNUAL AFFAIR

Self-assembly, modular planters can be made into many shapes
and are ideal for decorating an awkward corner or area of the
patio, especially when filled with bright annuals.

MATERIALS LIST
You will need: self-assembly modular planter (1); potting compost (2); moisture-retentive pellets (3); trowel (4); selection of Summer annuals (5).

S elf-assembly modular planters are practical and flexible. They are robust enough to form a permanent feature that you can tailor to your own design, but are also easily dismantled and stored so that they can be used for temporary plant displays.

A bright idea

This type of planter makes an ideal container for annuals, especially if you only wish to create a display for the Summer months. The bright white plastic has a smart appearance that complements pastel shades of the likes of Stocks, Lobelia, Petunias, Geraniums, and Sweet Alyssum.

Fine foliage

Smaller designs are also ideal for use indoors, filled with a lush display of foliage houseplants, perhaps brightened up with some seasonal flowering plants. These types of planter tend not to be watertight, however, so will have to be lined with a strong layer of plastic sheeting.

PRACTICAL POINTERS

Flexibility itself Although each kit will only consist of a few types of pieces, there are many ways in which they can be fitted together to form a vast array of shapes. Self-assembly planters of this type can usually be purchased in many different-sized kits; here are some ideas for the manner in which they can be used.

❑ **Long trough** Straight side panels can be fitted together to form a trough, as long as you require, with rounded or straight ends. This would be suitable for standing along a wall, or it could be used as a windowbox on a wide ledge.

❑ **Simple shapes** Countless simple shapes can be made, including a circle, a square, a heart shape, a hexagon, or even a 'cloud'.

❑ **Corner features** This system is ideal for filling a corner, making use of otherwise unusable space. Try an L-shaped design, or a triangular style for a larger feature.

❑ **Multi-tiered planters** Use the pieces to create any number of tiers, in any configuration, even covering a whole area of the patio, with sections at different heights.

❑ **Towers** Stack several modules on top of each other to form a tower that can be planted all the way up with a cascade of colour.

❑ **Tree-surround planter** Sections can be built around a tree-trunk to form a ready-made flower bed. Use the system without its feet for this.

ASSEMBLING THE PLANTER

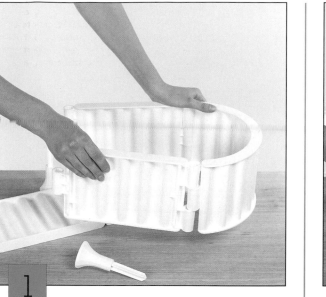

1

The self-assembly planter consists of straight, bowed and half-round side panels, base pieces, and various pins and feet to hold the panels together. The first step is to decide upon the shape you wish to make, then slide the panels together accordingly.

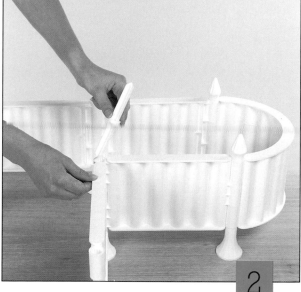

2

Use a pin with a decorative finial for the joints which will not be topped by another section of the planter. For those that are to be joined to a section above, use a double-ended pin, leaving one end protruding upwards. Attach the feet to the bottoms of the pins.

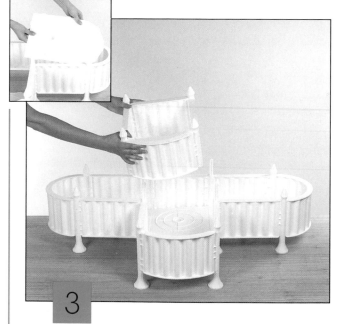

3

Assemble the base plates, using the joining strips supplied, then slide them into the planter and adjust until the base is flat and level (inset). If making a multi-level planter, assemble the upper section and fit it onto the bottom planter, using the protruding pins.

4

Before filling the planter with compost, put in a layer of drainage material, such as moisture-retentive pellets or crocks, spreading it out over the bases in an even layer. If you are using moisture-retentive pellets, make the layer about 2.5cm (1in) deep.

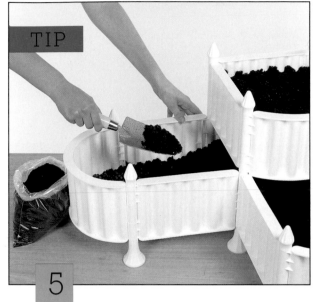

TIP

5

Most Summer annuals are best grown in a free-draining compost, so if necessary, add up to a quarter by volume of horticultural grit to improve the texture of the mixture. For other types of plants, confirm the compost requirements before planting them.

6

Having half-filled the sections with compost, plant-up the top section. Use quite tall, erect specimens, such as *Pelargonium,* in the centre and trailing plants, such as *Glechoma,* to cascade down over the sides. Trailing plants will soften the hard lines of the planter.

7

Use pretty annual Stocks to fill up the lower compartments, softening the edges with *Glechoma* and *Brachycome.* Sweet Alyssum is ideal for introducing a splash of cool white. Fill the spaces between the plants with more compost and firm lightly all over.

8

As soon as you have finished planting, water the plants thoroughly, using a watering can fitted with a fine rose. Moisten the compost all over and continue to water until drips appear out of the bottom. Try not to knock the plants over with the spray of water droplets.

DAINTY SNOWDROPS

Delicate, nodding Snowdrops, massed in a terracotta log planter, will create a quaint and dainty display to welcome visitors on a front doorstep in late Winter.

MATERIALS LIST
You will need: terracotta log (1); compost (2); bark chips (3); crocks (4); Snowdrops, *Galanthus* species (5); trowel (6).

Snowdrops are usually the very first of the Spring bulbs to emerge, their nodding white blooms and attractive fine foliage appearing early in the year, often from beneath a blanket of snow.

There are many forms of the Common Snowdrop (*Galanthus nivalis*), ranging from a beautiful double form (*G. n.* 'Flore Pleno') to hybrids with green-tipped petals, such as *G. n.* 'Pusey Green Tip', and even an unusual form with clear yellow markings, *G. n.* var. *lutescens*.

Plant the dormant bulbs around September, placing them in a semi-shade position. Keep the compost moist at all times. The Snowdrops will soon form clumps to produce more flowers each year. Lift and thin out the clumps every few years, planting the spare bulbs in another tub.

PRACTICAL POINTER

❑ **More flowers** The Snowdrops can also be planted 'in the green' in March. Plant the bulbs to the same depth as the dry bulbs and leave the foliage to die down. Using this method, they will probably become established more quickly, and may produce more flowers in the first year.

1

Place small crocks in the bottom of the terracotta log, making sure that you cover the drainage holes to stop compost from leaching out during watering. If there are no drainage holes, put in a good layer of crocks.

2

Use a free-draining, humous-rich compost in the log, adding a handful of leaf mould or well-rotted manure to the mixture if possible. Fill the log to a level 2.5cm (1in) below the rim, firming the compost as you go.

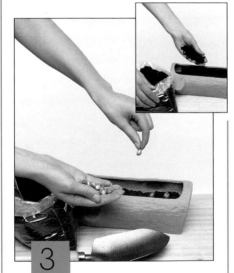

3

Arrange the bulbs quite close together, pointed ends upwards, on the surface of the compost. Sprinkle on more compost and firm between the bulbs to fill in air spaces. Fill the log to just below the rim (inset).

4

After watering the bulbs well, cover the surface of the compost with a mulch of bark chippings. Not only will this look attractive, but it will also serve to prevent the compost underneath from drying out too quickly.

HEAVENLY SCENT

Positioned by a doorway or path, a group of terracotta pots planted with scented Geraniums will bring delicate colour and a delicious, light fragrance to the patio.

PLANTING THE GERANIUMS

MATERIALS LIST
**You will need: terracotta pots (1);
Rose Geranium, *Pelargonium
graveolens* (2); Fern Leaf Geranium,
P. denticulatum 'Filicifolium' (3);
Lemon Geranium, *P.* 'Lady
Plymouth' (4); Nutmeg Geranium, *P.
x fragrans* (5); compost (6); crocks
(7); trowel (8).**

Scented Geraniums were a favourite with the Victorians, who used them indoors and out, where long skirts would brush against the foliage and scent the air with the fragrance of lemons, roses, apples, peppermints, nutmeg or balsam.

These plants not only smell nice, but look good too. Leaves may be crinkled or strongly dissected, often with a cream edge, and dainty pink flowers are produced in Summer.

For an eye-catching display, plant a group in terracotta pots and place them by a doorway or beside a path, where passers-by will brush against them. A hot, sunny position is best.

RECIPES & REMEDIES

❏ **Flowers** are delicious tossed into salads.
❏ **Leaves** can be used to flavour sauces, jellies and jams (they taste the same as they smell). Lay them under baked apples and cakes to impart flavour (remove before serving), or infuse as a tea for a whole range of different tastes. Use dry in pot pourri mixtures.

1

Place a good handful of crocks into the bottom of each terracotta pot. Make sure that the drainage holes in the bottoms of the pots are covered by the crocks to prevent compost from leaching out during watering.

2

Use a free-draining compost in the pots, adding up to a quarter by volume of horticultural grit to the compost, if necessary, to improve the texture. Fill each pot to a level about 2.5cm (1in) below the rim.

3

Remove the *Pelargonium* plants from their pots by turning upside down and tapping each pot on a firm surface to release the rootball. Take care not to damage the roots or crush the foliage.

4

Decide which of the plants will look best in which of the pots, taking leaf shape, size and general form into account. Position the plants in the chosen pots, firming compost gently back around to secure them.

A FOUNTAIN OF FOLIAGE

This classically-styled stone urn planted with a single superb Cordyline provides a striking architectural feature which will look impressive throughout the year.

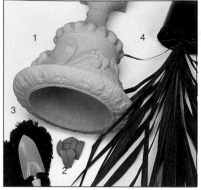

MATERIALS LIST
**Stone urn (1); crocks (2); compost
(3); *Cordyline australis* (4).**

Formal areas of garden need simple architectural features as a focal point, especially in Autumn and Winter when there's less interest from plants and flowers. A classic urn looks impressive when planted with a single evergreen shrub, such as *Cordyline australis*. The cultivar 'Purple Tower' has broad, sword-shaped leaves of intense purple-red growing in spraying habit. It looks striking planted near similar-coloured specimens such as Sedum 'Autumn Glory'. Hardy in a southerly aspect, it does benefit from being sited in a sheltered spot, protected by walls or hedges.

PRACTICAL POINTERS

❏ **Discolouration** If lower leaves wither or discolour, snip off neatly with scissors.
❏ **Plant with urn** in place; it will be heavy and unwieldy when finished.
❏ **New urns** made from reconstituted stone can be 'weathered' by applying diluted liquid manure or yoghurt to attract algae; proprietary treatments are also available.

1 Garden urns usually have a drainage outlet where top fits onto pedestal; if not, you should drill a hole. To ensure compost stays in place, put a small layer of crocks in bottom of urn.

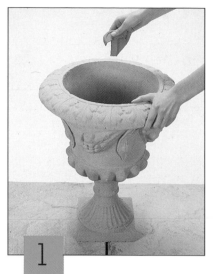

2 Fill the urn with a humus-rich compost suitable for potting-up shrubs. Do not over-fill, however; fill to a level which will allow you freedom to position the plant at right height.

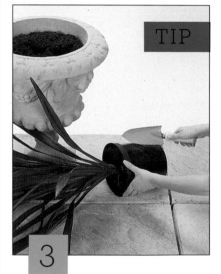

TIP

3 Lay the plant - still in its pot - on its side on the ground and tap the pot all round to loosen the rootball. Holding plant at base of stem, remove it, being careful not to damage delicate root system.

4 Place plant upright in urn, checking that it is at the right height. Trowel more compost around the plant, firming it down carefully to fill any gaps. Level compost and water-in well.

AUTUMN HUES

Rose-pink and white Heathers are ideal for an Autumn container; teamed up with rich blue conifers and trailing Ivies, they create a splash of colour when other plants are dying back for the Winter.

MATERIALS LIST
You will need: plastic rectangular tub (1); Lawson Cypress, *Chamaecyparis lawsoniana* **'Ellwood's Silver' (2); Heather,** *Erica gracilis* **(3),** *Calluna vulgaris* **'County Wicklow' (4), and** *Calluna vulgaris* **'Kinlochruel' (5); Ivy,** *Hedera helix* **'Lutzia' (6); ericaceous compost (7); trowel (8); crocks (9); bark chippings (10).**

Heathers are extremely versatile and popular plants and it is possible to buy varieties in flower at any time of the year, making them useful in many situations. There is a large choice of Autumn-flowering varieties which, when teamed-up with rich blue conifers and trailing Ivies, make an attractive show of colour on the patio at a time of the year when many other plants are dying back for the Winter.

The mixed foliage will look good all through the Winter, but you could replace the Heathers that have finished flowering with other varieties, which flower during the Winter months, thereby extending the flowering period.

Striking effects
Choose slow-growing, columnar conifers that will form a neat row at the back of the trough or windowbox. After a few years, if they start to get too big for the container, you can trim them down or replace them with smaller ones. Silvery-blue varieties, such as *Chamaecyparis lawsoniana* 'Ellwood's Silver', have a most striking effect when coupled with deep rose-pink and white Heathers.

Soft tones
The trough will need to be frost-proof for an Autumn and Winter display. The soft tones of terracotta are complemented by conifers and Heathers, but cheaper and lighter is terracotta-look plastic which can be just as effective. For a smarter look, try a white plastic trough filled with white Heathers and blue conifers.

Depending on the size of the plants, you will need four or five conifers, three or four large Heathers and four smaller Heathers for a 75cm (30in) long trough.

A sheltered position
The Heathers will perform best in a sheltered, South-facing position, out of cold winds: the sun will encourage them to flower. Some South African varieties, such as the Rose Heath (*Erica gracilis*), are not fully hardy; use tougher plants for exposed sites. When the flowers have faded, remove unsightly dead heads. Water the trough well to keep the compost moist at all times.

PICK & PLANT

There are many species of *Erica* sold in garden centres; these are the most suitable for a tub:

❏ *Erica carnea* - of which there are innumerable cultivars in shades of white, pink and purple. *E.c.* 'Queen Mary' has deep rose flowers in November, while *E.c.* 'Snow Queen' is a pure white variety which flowers in December.

❏ *Erica cinerea* (Bell Heather) - these cultivars flower in early Autumn. *E.c.* 'Alba Major' is a compact, white-flowering form, *E.c.* 'Ruby' has rose-purple flowers, while *E.c.* 'Sea Foam' has pale purple blooms.

❏ *Erica gracilis* (Rose Heath) - this is a South African species that is not fully hardy, but is grown for its vivid, strong cerise flower colour.

PREPARING THE TROUGH

First put some crocks in the bottom of the trough, making sure you cover the drainage holes to prevent compost leaching out during watering. If there are no holes in the bottom of the trough you will need to make some with a sharp knife.

Half fill the trough with ericaceous compost, leaving room on top for the plant rootballs. Ericaceous compost must be used because Heathers are lime-haters and will not tolerate normal potting compost, they need acidic conditions.

Remove the conifers from their pots and stand them in a tight row along the back of the trough, with the base of each plant coming about 1cm (1/2in) below the trough rim. Push them close together and aim to create a solid screen of foliage.

TIP

In front of the conifers, form a row of the large *Erica*, placing one in each space between two of the trees, leaving a gap at either end for the Calluna plants. If you have used five conifers, you will need four *Erica*; or three *Erica* with four conifers.

COMPLETING THE TROUGH

5 Make sure the bases of the plants are all at the same level, about 1cm (3/8in) below the rim of the trough; adding more compost if necessary. Fill in the spaces between the roots with compost, firming with your fingers between the plants.

6 Plant two of the Calluna plants at each end of the trough, encouraging the foliage to spill out over the sides to soften the edges. If you have two different colours of Calluna, plant one of each at both ends to create a symmetrical effect.

7 Fill in any spaces with Ivy plantlets. Usually there are about eight small plants in each pot bought from a garden centre; separate these and plant them individually along the front of the trough and down the sides, arranging them evenly.

8 Finish by firming the compost all around and adding more if there are any spaces. Water thoroughly and cover the compost with a layer of bark chips. This mulch looks attractive and will help to prevent the compost drying out.

CORNER PLANTS

Create a splash of year-round colour and lush foliage in a plain
entrance porch with a compact arrangement of evergreen plants
in a stone-look corner planter.

MATERIALS LIST
You will need: a log planter (1);
English Daisies, *Bellis perennis* **(2);**
crocks (3); compost (4); trowel (5).

The English Daisy is a charming, old-fashioned favourite that is closely related to the lawn Daisy. In the Spring, the fleshy rosettes of oval leaves throw up masses of pretty flowers in shades of red, pink and white. Single, double and pom-pom forms are all available, and they make an ideal display for a rustic log planter.

Plant-up the log in the Autumn; the plants will grow through the Winter and produce a good show of colour in the Spring. Keep the compost just moist in really cold weather. Place the log in a sunny, or semi-shade, position and dead-head regularly to prolong flowering.

PICK & PLANT

❏ **Following on** English Daisies are grown as biennials and, thus, are discarded in the Summer after flowering. Replace them with Summer bedding plants, such as Geraniums and Marigolds, and when these have finished in the Autumn, plant more Daisies for the following Spring.

Place several handfuls of crocks into the bottom of the log planter and spread them out along its length. Make sure you cover the drainage holes to prevent compost from leaching out during watering.

English Daisies require a very free-draining compost, so add about a quarter by volume of horticultural grit to the mixture to improve the texture. Fill the log to within 2.5cm (1in) of the rim, firming the compost as you go.

If the Daisies are in a polystyrene tray, remove them carefully by pushing up from underneath. Take care not to damage the roots. Plant the Daisies close together in the log for a really full effect in the Spring.

Lower the roots of each plant into a small hole made in the compost, then lightly firm compost back around them with your fingertips. Water the plants in well to help them become established quickly.

CABBAGE CREATION

A fountain-like Cabbage Palm, teamed with brightly-coloured Ornamental Kale plants, forms a delightful architectural display for the late Summer and Autumn months.

CREATING THE DISPLAY

MATERIALS LIST
You will need: Cabbage Palm,
Cordyline australis **'Doucetii' (1);**
trailing variegated Ivy, *Hedera helix*
(2); crocks (3); trowel (4); compost
(5); Ornamental Kale, *Brassica*
oleracea **(6); large tub (7).**

These unusual Ornamental Kale plants come into the shops in the late Summer and will give a cheerful display all Winter long. When combined with an arching Cabbage Palm and long trails of waving Ivy, the shapes and textures complement each other to form a dramatic and architectural centrepiece for a wall or patio.

Stand the tub in a bright position outside; the plants will not be damaged by strong winds, so an exposed site is fine.

When the risk of frost approaches, either split up the display and move the frost-tender Cabbage Palm to a sheltered place for the Winter, or move the whole tub into a cool greenhouse, or a cool conservatory or hall, where it can be appreciated all Winter long.

PRACTICAL POINTER

❏ **Heavy work** When the tub is planted up it will be extremely heavy, so it is advisable to put the display together in its final position. If you decide to move the display, intact, into a frost-free place for the Winter, either get someone strong to help you or use a sack trolley for the purpose.

1
Put plenty of large crocks into the bottom of the tub, making sure that they cover the drainage hole to prevent compost from leaching out (inset). Trowel-in compost on top, filling the tub to within 8cm (3in) of the rim, firming as you go.

2
Remove the Cabbage Palm from its pot by turning it upside down and tapping the pot on a firm surface to release the rootball. Plant it in the centre of the tub, at the same depth as in the pot, and firm compost around the rootball to secure it.

3
Plant the Kale around the edge of the tub, alternating the pink and cream colours. Arrange them so that the leaves are held up by the pot rim and don't rest on the damp compost, as they may rot (inset). Fill between them with compost.

4
Fill the gaps between the Kale plants with trailing Ivy. Pots of Ivy from garden centres usually contain about eight well-rooted cuttings; you may need to split the pots into two groups of plants to fit them between the Kale plants comfortably.

NASTURTIUM TOWER

A wrought-iron pot holder is perfect for displaying many plants, especially Summer annuals, and is ideal for exploiting the limited space on a small patio or terrace.

MATERIALS LIST
You will need: wrought-iron pot stand (1); selection of bedding plants, such as Nasturtium (2), Mimulus (3), Verbena (4), and Petunia (5); crocks (6); pots (7); trowel (8); horticultural grit (9); compost (10).

Small pots often get 'lost' on the floor of a patio, so it is a good idea to raise them closer to eye-level, where the plants in them can be appreciated more fully. A wrought-iron stand, which has special rings to hold pots, is ideal for this and offers an inexpensive means of displaying several plants as an attractive feature.

PICK & PLANT

A varied choice Many plants are suitable for this type of display:
❏ **Bedding plants** For variety of colour and seasonal interest, bedding is the obvious choice and there is a wide array of plants, including many miniature bulbs, from which to choose.
❏ **Alpines** A pot stand is ideal for displaying dainty alpines, which are less visible on the floor. Cushions of Saxifrage look really good in small pots.
❏ **Ivies** The pots could be filled with different Ivies, including some of the many variegated ones now available, for evergreen interest.
❏ **Hardy ferns** A collection of the smaller hardy ferns would be another good choice.

PLANTING THE HOLDER

1

Place crocks in the bottom of each pot to prevent the compost from being washed out during watering. Pieces of broken clay pot, or pieces of slate are ideal, but make sure they are not too large, or they will not fit in to the small pots.

2

Use a free-draining compost, adding up to a quarter by volume of horticultural grit to the mixture, if necessary, to improve the drainage (inset). Fill all the pots to about half full with the compost and firm lightly with your fingers.

3

Depending on the size of the pots, you will probably only need one or two plants in each. Team trailing plants with upright, bushy ones for a balanced effect. Carefully remove the plants from their pots to avoid root damage.

4

If the plants have become pot-bound, it is a good idea to tease out a few of the roots to encourage them to spread out into the available space. Fill the gaps around the roots in the new pots with more compost and firm gently.

A BRIGHT PALETTE

Plant a terracotta pot with bold Tulips and pure white bells of Heather for a short-term, but vibrant, splash of colour on the patio in Spring.

MATERIALS LIST
You will need: terracotta pot (1); Heather, *Erica carnea* 'Snow Queen' (2); Water Lily Tulip, *Tulipa kaufmanniana* 'Stresa' (3); compost (4); crocks (5); trowel (6).

This colourful arrangement of Tulips and Heathers requires careful selection of plants to ensure that they will be in bloom at the same time. Here, Tulip 'Stresa', with its stripy foliage and vibrant yellow and scarlet blooms, is effectively offset by the pure white bells of Heather 'Snow Queen'. However, many other combinations would be just as eye-catching, so spend some time browsing through bulb and plant catalogues or the garden centre.

A splash of colour
This display is designed to provide a short, but very bright, splash of colour for the Spring. When the Tulips have finished flowering, replace them with Summer bedding plants to complement the rich foliage of the Heathers.

For a cheerful Spring display, plant-up the container between October and December. When the Tulip flowers have faded, remove the bulbs and place in the warm to dry; when the foliage is crisp, remove it and store the bulbs somewhere cool until the Autumn. Then they can be put back in the pot, or planted in a border to flower next year.

1

Put a good handful of large crocks into the bottom of the pot, covering the drainage hole (inset). Fill the pot with a good-quality, peat-based compost, to within 13cm (5in) of the rim, and firm gently with your fingers.

2

Arrange the bulbs on top of the compost, pointed ends upwards, between 3 and 5cm (1 and 2in) apart. If the rim of the bowl overhangs the compost, place the bulbs so that they can grow up straight.

3

Sprinkle compost on top of and between the bulbs, making sure that there are no gaps. When the bulbs are covered, firm gently all over, then continue adding compost up to 2.5cm (1in) below the container's rim.

4

Remove the Heather plants from their pots by turning them upside down and tapping each pot to release the rootball. Plant the Heathers, equally spaced, around the edges of the pot. Water the display well.

JEWEL-BRIGHT HEATHERS

Heathers are not only among the easiest plants to grow but also provide a brilliantly-colourful mass of foliage and flowers throughout the year, when planted in a simple terracotta bowl.

PLANTING THE HEATHERS

MATERIALS LIST
The display consists of a deep, round unglazed clay bowl, ericaceous compost, crocks and clay pellets for drainage and a selection of heathers.

Evocative of the rugged, breathtaking beauty of open moorland, heathers (*Erica*) can make ideal container plants for the patio, in bright, even quite exposed locations.

This variety, Cape Heath, or Christmas Heather (*Erica gracilis*) has a cluster of brilliant rich pink or white bell-shaped flowers on spiky stems which, when massed together, will produce a glorious flush of brilliant colour.

Heathers can be used all year round: some have foliage that changes from red, yellow, orange and bronze to greens, silvery-greys and pinks. Autumn- and Winter-flowering heathers add interest when other plants die back, while Summer varieties assume a subdued colour when their flower heads have faded.

PRACTICAL POINTER

❏ **Soil Selection:** Heathers thrive in peaty, acid soil. Most, apart from more tolerant Winter-flowering types, need lime-free soil. For a container-grown display, choose pre-bagged ericaceous compost, formualated for lime-hating members of this family.

1

The container should have a drainage hole in the base. Cover with a piece of broken pottery then add 5cm (2in) layer of moisture-retentive clay pellets to assure proper drainage (inset).

2

Fill the container to the two-thirds mark with a specially-formulated ericaceous potting compost: an ideal medium for lime-hating plants such as heathers. Then firm the compost lightly.

3

Starting at the centre of the container, dig a hole in the compost and plant a single heather, starting with a white or pale pink type. Firm the compost around the root ball so that the heather is held upright.

TIP

4

Plant the remaining heathers in an ever-increasing spiral to fill the container. Space the heathers about 12.5cm (5in) apart so that they eventually fill out to form a dense, spiky mass of colour.

EXOTIC PERSIAN BUTTERCUPS

Raised from tubers, or bought as mature plants for an instant display, Persian Buttercups provide a mass of showy, colourful blooms, which are offset perfectly by dark blue ceramic pots.

PLANTING THE POTS

MATERIALS LIST
You will need: Persian Buttercups,
Ranunculus asiaticus **(1); decorative**
ceramic pots (2); trowel (3); peat-
based compost (4); crocks (5).

The showy, peony-like flowers of the Persian Buttercup come in a wide range of strong colours, which are perfectly offset by the dark blue of the ceramic pots used here. The result is a shock of vibrant colour which will cheer up a dull patch on the patio or in the garden.

Persian Buttercups are available, as mature plants from about March onwards, and are ideal for creating an instant Spring or Summer display. However, the plants will require frost protection in severe weather if used in early Spring. Place the pots in a sheltered, sunny position, and keep well-watered, allowing the compost surface to dry between waterings.

PRACTICAL POINTER

Inexpensive alternative
Persian Buttercups can be raised, far more cheaply, from dry tubers, which should be soaked in water overnight before planting. For Spring flowers, plant in late Autumn and protect from Winter frosts; for Summer blooms, start the tubers in a frost-free place in Spring, then move to their flowering positions at the end of May.

1 Place a good handful of crocks into the bottom of each pot, making sure you cover the drainage hole (inset). Then fill the pot to within 10cm (4in) of the rim, using a rich, peat-based compost and firming lightly as you go.

2 The soft foliage of the plants is liable to become tatty, especially if they are bought in leaf and are squashed into boxes for carriage. Check each plant in turn, removing any bruised or broken leaves.

3 Gently remove the plants from their pots, taking care not to squash any remaining foliage. Stand them in the ceramic pot to gauge their positions. If using a mixture of colours, distribute them evenly.

4 For the best effect, mass the plants close together in the pots; finish by filling the gaps between the roots with fresh compost, firming it gently between them. Water the plants well, moistening the compost thoroughly.

CHILD'S PLAY

A plain wooden window box, filled with cheerful Pansies, creates a splash of Winter colour for a Wendy house and is simple enough for children to plant for themselves.

MATERIALS LIST
You will need: Winter-flowering Pansies, *Viola x wittrockiana* **(1); wooden window box (2); compost (3); trowel (4); moisture-retentive pellets (5); paint brush (6); preservative (7).**

Choose a bold mixture of colours to make the box cheerful and appealing to children; don't worry about the colours clashing, the brighter the better.

The plants will be available in garden centres in the Autumn, and will flower all Winter long, especially if you remember to remove the dead flower heads. You will need six or eight Pansies for a 75cm (30in) long trough, depending on the size of the plants; if there are any left, make another window box or a Pansy hanging basket to match.

PRACTICAL POINTER

❑ **Preservation** For outdoor use, wood must be treated with a preservative to prevent it from rotting. Follow the manufacturer's instructions and let the box dry overnight before planting. Children must be supervised if attempting this themselves.

1

If the window box has no drainage holes, pour a 3cm (1in) layer of moisture-retentive pellets into the bottom to soak-up excess moisture and prevent root rot. If the box has holes, cover them with a few crocks.

2

Using a free-draining compost, fill the box almost to the top, firming as you go and leaving room for the rootballs. Make sure the compost goes into all the corners and is the same depth along the length of the box.

3

Remove the plants from their pots by tapping each pot with the trowel to release the rootball, then lift the plant out gently. Remove a little compost from the planting position, lower in the rootball and firm around it.

4

Repeat with all the plants, spacing them about 8-13cm (3-5in) apart, depending on the size of the plants, until the box is full. Aim for an even spread of colours along the length of the box. Water the plants in, using a rose on the watering can.

INDEX